Quantum Big Bang-Quantum Vacuum Universes (Particles)

A Varying Hubble Constant with a Big Dip
Dark Energy Dominance
New QED-like Vector Interaction
Universes' Eigenvalue Function
Semi-Free Universe Particles
Proof: Vacuum Polarization "Expands" Universe
Voids and Superclusters Mechanism
Universal Scale Factor for Expansion/Contraction

Stephen Blaha Ph. D.
Blaha Research

Pingree-Hill Publishing
MMXIX

Cover: The cover depicts the time varying Hubble Constant, and corresponding universe expansion and contraction.

Rev. 00/00/01   August 21, 2019

To John, Natasha, and Stephen

Some Other Books by Stephen Blaha

All the Megaverse! Starships Exploring the Endless Universes of the Cosmos using the Baryonic Force (Blaha Research, Auburn, NH, 2014)

SuperCivilizations: Civilizations as Superorganisms (McMann-Fisher Publishing, Auburn, NH, 2010)

All the Universe! Faster Than Light Tachyon Quark Starships & Particle Accelerators with the LHC as a Prototype Starship Drive Scientific Edition (Pingree-Hill Publishing, Auburn, NH, 2011).

Cosmos Creation: The Unified SuperStandard Model, Volume 2, SECOND EDITION (Pingree Hill Publishing, Auburn, NH, 2018).

Immortal Eye: God Theory: Second Edition (Pingree Hill Publishing, Auburn, NH, 2018).

Unification of God Theory and Unified SuperStandard Model THIRD EDITION (Pingree Hill Publishing, Auburn, NH, 2018).

Calculation of: QED α = 1/137, and Other Coupling Constants of theUnified SuperStandard Theory (Pingree Hill Publishing, Auburn, NH, 2019).

Coupling Constants of the Unified SuperStandard Theory SECOND EDITION: We Find the Fine Structure Constant 1/137.0359801, and so: OUR UNIVERSE AND LIFE! Also a Universal Eigenvalue Function for all Known Interactions, And Running Coupling Constants to all Perturbative Orders (Pingree Hill Publishing, Auburn, NH, 2019).

Available on Amazon.com, bn.com Amazon.co.uk and other international web sites as well as at better bookstores (through Ingram Distributors).

CONTENTS

FIGURES and TABLES

INTRODUCTION

Two of the major problems of Cosmology are the nature of the Big Bang and the process of expansion of the universe. In this book we consider our Quantum Big Bang model in more detail and show its solution eliminating divergences in previous Big Bang models. We also consider our universal scale factor a(t) in great detail focusing particularly on the large Dark Energy that it implies. The scale of the Dark Energy is much more than past very large estimates of quantum vacuum energy.

Progressing further we create a quantum theory of a vector QED-like interaction within, and between universes. Calculating the universe's vacuum polarization using the same techniques that led us to calculate the QED fine structure constant successfully we calculate the interaction coupling constant, and the resulting vacuum polarization of the universe. Remarkably the fourier transform of the vacuum polarization is our universal scale factor for the universe's expansion. Thus we obtain a deeper meaning for the universe's expansion over time. It is governed by the universe's vacuum polarization.

Since the new vector interaction must be the same for any universe in the cosmos we find that all universes have the same evolutionary pattern as ours. Further, we find that an eigenvalue condition exists (and is satisfied) for our universe. It implies that our universe is a quasi-free particle (gravitation makes it quasi-free) in the cosmos that we have called the Megaverse. We describe our universe particle theory in some detail.

The Hubble Constant, which is a source of some controversy, is explained in detail as a function of time, as is the Dark Energy, universe density, universe pressure, and related quantities. We show, in more detail, the universe has a Big Dip, during which it contracts to 70% of its size. The existence of the Big Dip is necessary if we consider the known time dependence of the Hubble Constant.

The derivation and the consequences of our universal scale factor are described in detail with many graphs of the Hubble Constant, energy density, Dark Energy, pressure, equation of state, deceleration parameter, and other quantities of interest.

SECTION 1: MEGAVERSE PRELIMINARIES

1. Evidence for Entities Beyond Our Universe

1.1 Theoretical Evidence for Other Universes and External Matter

Why are we not content with one universe given its enormous size and variety? It appears that there are important theoretical reasons, and some important experimental observations, that suggest that there is more than our universe 'out there.' The external entities are likely to be other universes but external "clumps" of mass-energy are not excluded.

In this chapter[1] we will discuss theoretical reasons and experimental suggestions of a larger space—that we call the *Megaverse*—that contains our universe and, most likely, other universes. The existence of a Megaverse resolves important theoretical issues and may resolve some important astronomical puzzles that have appeared in recent years.

The theoretical issues, which have been subjects of discussion for some time, are:

1. The need for an external 'clock' to measure 'time' knowing that it is to some extent relative and local.
2. The need for a 'quantum observer' to complete the understanding of quantum gravity as described by the Wheeler-DeWitt equation and other efforts to develop a quantum gravity.
3. The need for other universes to provide theoretical measuring platforms for quantities beyond the charge and mass-energy of the universe. We think here of the other quantum numbers of particles and particle number operators such as Baryon number.
4. The need for an ultimate source of mass and inertia in our universe.

---

[1] Most of this chapter appears in Blaha (2015a) and in earlier books by the author.

In Blaha (2015a), and earlier books, we have suggested that there are weighty reasons to believe that other universes exist.[2] The existence of other universes solves these problems.

These problems generally have a source in Quantum Gravity and the interpretation of the Wheeler-DeWitt equation in particular. See Blaha (2017c) and (2018e) for discussions of the Wheeler-DeWitt equation and its implications. We now consider the above issues.

1.1.1 Universe Clocks

Asynchronous Logic provides the equivalent of a clock for the synchronization of processes within large electrical systems such as VLSI chips. Similarly there is a need for a universal clock for our universe. As DeWitt[3] points out in his studies of quantum gravity,

'"The variables … [of the quantized Friedmann model] because of their lack of hermiticity, are not rigorously observable and hence cannot yield a measure of proper time which is valid under all circumstances. … . It is for this reason that we may say that "time" is only a phenomenological concept … If the principle of general covariance is truly valid then the quantum mechanics of everyday usage with its dependence on the Schrödinger equations … is only a phenomenological theory. For the only "time" which a covariant theory can admit is an intrinsic time defined by the contents of the universe itself. Any intrinsically defined time is necessarily non-Hermitean, which is equivalent to saying that there exists no clock, whether geometrical or material, which can yield a measure of time which is operationally valid under *all* circumstances, and hence there exists no operational method for determining the Schrödinger state function with arbitrarily high precision."

---

[2] In Blaha (2013a), before the Higgs particle was discovered at CERN we suggested an alternate mechanism was possible if a sister universe existed (making the existence of other universes a reasonable possibility. The Higgs discovery makes the sister universe mechanism unlikely.

[3] DeWitt, B. S., Phys. Rev. **160**, 1113 (1987).

The lack of a clock within our universe invalidates quantum mechanics in principle and Quantum Gravity in particular. DeWitt concludes, "Thus [quantum gravity] will say nothing about time unless a clock to measure time is provided." And it appears absolutely necessary.

Unruh[4] also has an issue with the source of time:

"One of the key problems is that of time. We see and experience the world in terms of time. We see things grow, develop, and change. However, time does not enter into the Euclidean formulation of quantum gravity directly. In the usual Hamiltonian formulation, the Hamiltonian for quantum gravity is made up of densities which are the generators, not only of spatial coordinate transformations, but also of temporal coordinate transformations. The content of four of Einstein's equations is that some generators are zero. Thus all wave functions are invariant under all spatial and all temporal coordinate transformations. There is nothing in the wave function or the amplitudes which refers to the coordinate t, or the corresponding points of the manifold in any way. How then do we recover the indubitable and ubiquitous experience we have of time? The standard answer is that our experience of time is actually an experience of different correlations between physical quantities in the world. Time is replaced by the readings of clocks. I know that time has changed, not through any direct experience with time, but because the hands of my watch have changed.

Although the implementation of this idea is actually extremely difficult in practice, and although I personally believe that one should formulate one's quantum theory of gravity so as to contain time explicitly, let us nevertheless pursue the consequences of this idea of time as defined internally, as the "reading" of a dynamic variable. For an observer inside the theory, his "time" is not the coordinate; rather his time is some one of the given dynamic variables of the theory: y or P. Thus although the coupling to the baby universes via the effective action S is independent of the coordinates t or x, that does not mean that the observer inside the theory will experience the interactions as being independent of time. For him and/or her, time is one of the dynamic variables and so it can depend on the various dynamic variables of the theory,

---

[4] Unruh, W. G., Phys. Rev. D **40**, 1053 (1989).

even if it does not depend on the time coordinate t. In general one would expect the observer to see what looks to him like a time-dependent interaction with the baby universes. At one time, some one of the baby universes may couple strongly to the large universe, while at some other time, another of the baby universes will couple more strongly."

In Blaha (2015a) and earlier books, we suggested the existence of other universes provides a 'clock mechanism' in principle for our universe. And being universes themselves, these other universes are excellent clocks. DeWitt points out,

"Because every clock has a "one-sided" energy spectrum, its ultimate accuracy must necessarily be inversely proportional to its rest mass. When the whole universe is cast in the role of a clock, the concept of time can of course be made fantastically accurate (at least in principle) … "

Setting a mass scale using other universes, also sets[5] a time scale and resolves the issue of a clock for our universe. *In principle the existence of other universes validates the role of time in the Copenhagen interpretation of Quantum Mechanics.*

1.1.2 Quantum Observer

Attempts to create a quantum gravity theory have to confront the need for an *Observer* in any quantum theory within the context of the Copenhagen interpretation. DeWitt points out,

"The Copenhagen view depends on the assumed a priori existence of a classical level to which all questions of observation may ultimately be referred. Here, however, the whole universe is the object of inspection; there is no classical vantage point, and hence the interpretation question must be re-argued from the beginning. While we do not wish to stress this point unduly, since, after all, the Friedmann model ignores the

---

[5] For example the Planck time value is set by the Planck mass.

vast complexities of the real universe, it is nevertheless clear that the quantum theory of space-time must ultimately force a deviation from the traditional Copenhagen doctrine."

And Unruh states

"One of the key features in the interpretation of such transition amplitudes, or wave functions, is the idea that we, as observers are also a part of the Universe as a whole. We, as physical observers, must be describable from within the theory and not as observers external to the theory as in usual quantum mechanics. In usual quantum mechanics, the interpretation is usually given in terms of observers that are outside of the theory. There one makes a split, with the quantum world at one side of the split, and the observer on the other. von Neumann argued that the predictions of quantum mechanics, at least under certain assumptions, are independent of the exact location of that split, but Bohr argued adamantly for the necessity of such a split (classical observers and quantum world). *There is a great difficulty in setting up such a split for physical observers contained within and influenced by a quantum universe,* [italics added] and for the Universe as a whole, especially including gravity, one cannot argue that the predictions will be independent of where one puts the split. Since all energies interact gravitationally, and our observations are surely energetic phenomenon, the treatment of the energetics of observation as classical would lead to different predictions than if they were treated quantum mechanically. One is therefore forced to devise an interpretation of quantum mechanics in which the observer is part of the quantum system, rather than outside the quantum system.

This means that the interpretation of these transition amplitudes becomes somewhat non-intuitive. One must ask what the system looks like from within, from the viewpoint of an observer who is part of that world, rather than being able to interpret them directly in terms of probabilities for observations made by an external observer."

While the *Observer* question is addressed by a number of authors, the proposed answers are not entirely convincing. *The existence of other universes provides macroscopic Quantum Observers for our universe.* And our universe acts as a macroscopic quantum observer for other universes. Thus the quantum observer issue is resolved within the Megaverse of universes.

These considerations lead us to view the existence of other universes as a critical solution to the above problems.

1.1.3 The Higgs Mechanism is Explainable by Extra Dimensions

The Higgs Mechanism 'explains' (generates) fermion and boson masses. However the Higgs potential contains a quadratic term with a constant with the dimensions of [mass]. In a sense the Higgs Mechanism trades one mass for another. From where do the Higgs potentials' masses come?

A further explanation is needed is to determine the origin of the "dimensionful" mass terms in the Higgs' particle equations themselves. At present little if any thought has been given to the origin of these terms. We suggested that, excluding a "deus ex machina" source, the only known way to generate these mass terms in the Higgs' equations is through the separation of equations technique of differential equations. This technique requires additional parameters which can only be the coordinates of *extra unknown dimensions*. The best example of the generation of mass terms appears in the Schwarzschild solution of General Relativity where a separation constant, often denoted M, appears that has the dimension of [mass].

Thus extra space-time dimensions would resolve the origin of Higgs potentials' masses. Given extra dimensions it is reasonable to expect that these extra dimensions contain universes. Thus the Megaverse!

1.1.4 Possible Accretion of Megaverse Matter to Fuel Expansion of Our Universe

If matter is distributed outside of universes in the Megaverse, and if this matter can be accreted to universes by gravitational attraction, then the apparent increasing expansion of our universe may be due to this accretion. In chapter 14 of Blaha (2017c) we presented a model in which this possibility is realized. If true, then we would have tangible evidence of the existence of other universes in the Megaverse.

1.1.5 Asynchronous Logic is a Requirement of Universes

By establishing Asynchronous Logic principles[6] as the basis for the existence of universes and for setting the number of dimensions in each universe – four; and establishing the basis of fermion particles as *qubes*—we have found deeper principles of organization for the foundations of physics. The principles built on this foundation serve to enable the coordination of complex physical processes.

Usually we look at particle processes primarily from a space-time perspective: particles collide and produce new particles. We primarily think of the incoming and outgoing particles in a collision. However, considering the set of fundamental particles – and particle transforming interactions in themselves, while neglecting space-time and momentum considerations, leads us to view particles as constituting an alphabet and to view their interactions as a type of computer grammar.[7] Then the Asynchronicity Principles enable us to bring in space-time in a way that gives us the maximum complexity with the most minimal assumptions. As Leibniz[8] points out our universe has maximal complexity with minimal assumptions.

1.1.6 The Meaning of Total Quantities of a Universe

The 'external' properties of a universe are normally questioned—for the simple reason that it is assumed that there is no 'outside' of our universe. For example, Misner (1973) asserts:[9]

'There is no such thing as "the energy (or angular momentum, or charge) of a closed universe," according to general relativity, and this for a simple reason. To weigh something one needs a platform on which to stand to do the weighing."

---

[6] The basis of this section is described in detail in Blaha (2015a). That book places Physics within a logical framework that is a possible deeper ground for fundamental Physics theory.

[7] This conceptual approach was first described in Blaha (1998) who went on to characterize our universe as one enormous word evolving in time.

[8] See Rescher (1967).

[9] Pp. 457 - 458.

Misner et al presumes no such platform exists. If there is but one closed universe as most physicists currently believe then one cannot measure any totals of a closed universe such as ours. Yet if we take a more general view that our universe is only one of many then it becomes possible to measure total mass, charge, angular momentum, baryon number, and many other quantities of interest. Indeed, the existence of other universes (within the encompassing Megaverse) opens the door to an understanding of time, mass, energy, and all the other quantities necessary to develop a dynamical theory of universes.

Our new 'rotations of interactions' formalism (described in previous books) enables us to rotate measurable quantities. These quantities (quantum numbers) furnish a set of totals for our universe such as baryon numbers (normal and Dark), lepton numbers, angular momentum and so on that characterize our universe.

Later we will also see that one can then treat universes as 'particles', and develop 'universe dynamics', which might explain knotty problems such as the Big Bang and its precursor (if any). We will do this in subsequent chapters after first considering the possible structure of universes in general in the Megaverse.

1.2 Possible Experimental Evidence for the Megaverse

At first glance it would seem impossible to produce evidence for the existence of other universes. However there are subtle means by which we can 'sense' experimentally 'nearby' universes should they exist. The mechanism would appear to be gravitational effects exerted on objects within our universe by unseen nearby objects of enormous mass. Currently there appears to be three experimental suggestions for the existence of 'nearby' universes and one theoretical argument based on an influx of mass-energy from the Megaverse that may support an understanding of the expansion of our universe.

1.2.1 Great Attractors

One potential support is the discovery of the Great Attractor (at the center of the Laniakea Galaxy Supercluster), and the more massive Shapley Attractor (centered in the

Shapley Supercluster)[10]. These attractors contain massive numbers of galaxies and are drawing galaxies over a distance of millions of light years towards them.

If another universe(s) is 'near' our universe it could act as a 'gravitational magnet' and draw galaxies within our universe towards it to form one or more superclusters which could then act as attractors. Thus attractors might indirectly reveal the presence of other nearby universes—contrary to the expected large scale uniformity of the universe. The only other apparent source of superclusters is chance. Chance seems an unsatisfactory possibility in the present case.

1.2.2 Bright Bumps in Universe Suggesting Collision with Another Universe

A recent study[11] of the residual brightness of parts of the accessible universe found that bright patches appeared if a model of the CMB (Cosmic Microwave Background) with gases, stars and dust was 'subtracted' from the PLANCK map of the entire sky. After the subtraction one would expect only noise spread throughout the sky. However, bright patches were seen in a certain range of frequencies. These anomalies are thought to be a result of our universe colliding with another object – presumably another universe in the Megaverse.

1.2.3 Cold Spot in Universe Suggesting Collision with Another Universe

Another recent study[12] of a huge cold region of the universe spanning billions of light years revealed that this region is not a relatively empty region but rather is similar to in its distribution of galaxies to the rest of the universe. Previous the Cold Spot was considered an area where cosmic microwave background radiation – the leftover Big Bang radiation is weak – making it significantly colder (0.00015C colder) than the average temperature of the universe.

An analysis of 7,000 galaxy redshifts using new high-resolution data has now shown that the Cold Spot is similar to the rest of the universe. The Durham University

---

[10] Tully, R. Brent; Courtois, Helene; Hoffman, Yehuda; Pomarède, Daniel, "The Laniakea Supercluster of galaxies". Nature (4 September 2014). 513 (7516): 71–73; arXiv:1409.0880.

[11] Ranga-Ram Chary, arXiv.org:/1510.00126 (2015).

[12] T. Shanks et al, Durham University (Australia), Monthly Notices of the Royal Astronomical Society, 2016.

group suggested that the Cold Spot might have been caused by a collision between our universe and another Universe. They further suggested that there is only a 1 in 50 chance that it could explain this feature with standard cosmology.

Thus we have another important piece of circumstantial evidence in favor of other universes and thus the Megaverse.

1.2.4 Megaverse Energy-Matter Infusion into Our Universe

In chapter 14 of Blaha (2017c) we presented a model for an influx of mass-energy from the Megaverse to support the Bondi-Gold-Hoyle-Narlikar Steady State Cosmology, which was originally based on the 'continuous creation of mass-energy' by Hoyle and Narlikar. This model explains why the value of Ω makes the universe close to flat. If this model is correct then we would have concrete support for a Megaverse with a low mass-energy density leaking mass-energy into our universe. *More generally, it suggests that universes are surfaces of high mass-energy density in a Megaverse of low mass-energy density – with a ratio of mass-energy densities of the other of 10^{30}.*

1.2.5 Conclusion

We conclude that data is beginning to emerge favoring multiple universes and a physical Megaverse in support of the theoretical justifications presented earlier.

1.3 Historical Trend Toward Larger Space-Time Structures

Looking back through the history of Mankind's view of the universe we see a clear progression to a larger and larger view. Before the 16th century the earth was the universe. In the 16th century Giordano Bruno (and possibly others) suggested that the stars were suns with many worlds circling them. So our view of the universe expanded to include stars. Now we are on the threshold of external universes.

1.4 Philosophical Implications

Giordano Bruno (1548-1600) suggested the Many Worlds Hypothesis – there were many worlds in addition to the earth. This proposal led to a reassment of the place of Mankind in the Cosmos. The earth was no longer the only world, and Mankind was not necessarily the only intelligent species in the universe.

Now, with the possibility of many universes in the Megaverse, our universe becomes "not so special" and the prospect of innumerable worlds, possibly supporting other intelligent life, becomes likely. The religious implications of this new view are profound. They lead to a democratic perspective that Mankind is but one on many intelligent species and not necessarily of particular consequence to a deity.

Since the focus of this book is Physics we will not consider these issues here.

1.5 Possibly Universe Fragments

In addition to other universes the possibility of diffuse mass-energy in the Megaverse appears likely. The possibility of "mini-universes" is also likely. We leave that to future work.

2. Megaverse Dimension

Having seen strong motivation for the existence of a space of spaces (the Megaverse) we now turn to estimating its features. The determination of the dimensions of the Megaverse can only be described as guesswork unless a sound principle is consistently used to specify the dimensions. In this chapter we will use the known interactions (and the equal number of fermions) in our universe[13] to determine the Megaverse's dimension.[14]

We will assume that interactions have a dual role in fundamental physics: they determine the dynamics of particles, and they act to determine the dimensions of the universes and the Megaverse. The first role is evident within the universe and is therefore obvious from experiment. The second role is external to our universe. It acts to determine the dimensions of the Megaverse.

2.1 Role of Interactions in Determining the Megaverse Dimension D

The motivation for the second role can be discerned from considering a 2-dimensional space, and introducing a simple $1/r$ potential such as:

$$V = g^2/(x^2 + y^2)^{-\frac{1}{2}}$$

where g is a coupling constant.

One views V as the potential of a force. However the values of V suitably extended to the range $[-\infty, +\infty]$ can be viewed as a third dimension.

With this alternate perspective in mind we take the subset of Unified SuperStandard Model vector interactions,

---

[13] We use the Unified SuperStandard Theory to determine these quantities. See Blaha (2018e).

[14] Our general motivation for this choice is the relation between space-time and particles: without particles the measurement of distance in space-time is meaningless; without space-time particle dynamics is meaningless. The concepts of space and particles are "yoked" together. SuperSymmetry considerations somewhat support this view.

$$E = [SU(3) \otimes SU(2) \otimes U(1) \otimes SU(2) \otimes U(1) \otimes U(4) \otimes U(4)]^4$$

which has 192 generators, to suggest the Megaverse dimension is[15]

$$D = 192 \qquad\qquad (2.1)$$

It is important to note that the fields of the Unified Standard Model extend beyond universe boundaries since each point of a universe within a higher dimension Megaverse is surrounded by an infinity of Megaverse points. This feature is described in detail in the following chapter. Under these circumstances we require all fields be equally describable by the coordinates of the universe *and* Megaverse coordinates. Then continuity requires smooth transitions between values at universe points and Megaverse points in the neighborhood of universe points. Thus fields in our universe are intimately connected to their Megaverse counterparts, partially justifying our determination of the Megaverse dimension above. We provide further support for eq. 2.1 when we consider the 'rotation of symmetries' U(192) Θ-Group. (See Blaha (2018e.)

Lastly, we note for consistency, the Megaverse must have complex-valued coordinates since it encompasses the complex-valued coordinates of our universe (although a more complicated embedding can be visualized.) We opt for simplicity. In addition the metric of the Megaverse must be Euclidean with a flat-space metric tensor limit consisting entirely of –1's along the diagonal (using our conventions). For good reason we view the Megaverse as having at best a very low mass-energy density compared to that of our universe. Thus the Megaverse is essentially flat with 'universe bubbles.'

---

[15] We will use the symbol D to denote the dimension of the Megaverse in the following chapters.

3. The Embedding of a Universe in a Higher Dimension Megaverse – Surface Tension

3.1 Universes as Mass-Energy Islands

In developing the theory of the Megaverse we view universes as islands of mass-energy that maintain their 'integrity' as surfaces due to gravitation. Gravity holds universes together rather like molecular forces within a water droplet hold water molecules together. Molecular attractive forces give droplets cohesion as they (perhaps) descend through the earth's atmosphere. They are the origin of *surface tension* in water.

Similarly gravity holds higher density[16] mass-energy universes together and gives rise to gravity surface tension. In a model presented in Blaha (2017c) for the Big Bang and the expansion of the universe within the Megaverse we found the mass-energy density of the universe was a factor of 10^{30} more than the density in the surrounding Megaverse space.

Thus we have good reason to study the surface tension of universes as a result of gravitational attraction within universes.

3.2 Boundary of a Universe within the Megaverse

In this chapter we describe the embedding of universes within the Megaverse. Much of this chapter appears in several earlier books by the author such as Blaha (2015a).

As stated earlier, we define a universe to be a closed or open surface in the Megaverse with a much higher mass-energy density than the Megaverse.

There are two types of boundaries for a universe embedded in a space of larger dimensions. First there is a boundary of the universe determined by treating the universe as a surface in the space. Secondly, there is another type of universe boundary defined

---

[16] Higher density in comparison to the much lower density of the inter-universe space of the Megaverse.

by the observation that there are neighborhoods of every point that contain Megaverse points. Points of our universe are immersed in an inaccessible Megaverse sea of points. The cause is the higher dimensionality of the Megaverse.[17]

A point of our universe has a neighborhood with an infinite number of points of the enclosing Megaverse space.[18] Every universe point has neighborhoods with Megaverse points within it. Thus *each point of a universe is on a boundary of the universe due to the larger dimensions of the Megaverse space* within which it resides. Fig. 3.1 schematically illustrates these neighborhoods for any universe point for a universe contained within a higher dimensional Megaverse.

3.3 Confinement of Universes due to 'Surface Tension'

We will assume that other universes have the same physics as our universe with the possible differences that they may have differing interaction coupling constants and differing particle masses. As we stated above, every point of a universe in a higher dimensional Megaverse has Megaverse points in any neighborhood of the point (with the exception of neighborhoods strictly within the universe). Thus we confront the question: what keeps mass-energy at points in a universe, or is there leakage from the universe into the Megaverse?

If there is little or no leakage into the Megaverse then, since each point in a universe is part of a Megaverse surface, one can only assume that there is a barrier to movement into the Megaverse. Taking a note from fluid dynamics, and viewing Megaverse space as one 'material' and the universe as a different 'material,'[19] we view

---

[17] This point is illustrated by the following example: Consider a 3-sphere in 3-dimensional space. Points *within* the sphere do not have neighborhoods containing points outside the sphere. However all the points of a 3-sphere in a higher dimensional space have neighborhoods containing points with coordinate values of the other dimensions of the higher dimensional space. A simpler illustration is to consider a 2-dimensional (x, y) disk in 3-space. Each point of the disk has neighborhoods with points having a non-zero z coordinate.

[18] Any neighborhood of any point in the universe – with all its points strictly within the universe – has an infinite number of points within the universe. We assume the neighborhood is so small that the curvature of the universe's space can be neglected.

[19] Meaning material with much higher mass-energy density and consequently larger internal gravitational attraction.

the barrier as 'surface tension.'[20] The Megaverse appears to "exert a force" confining the contents of the universe to within itself.[21]

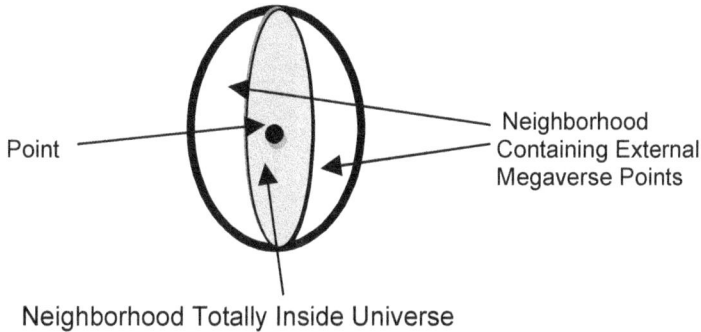

Figure 3.1. Schematic diagram of a 3-dimensional projection of 'orthogonal' neighborhoods of a point within a universe with one neighborhood strictly within the universe and the other neighborhood containing external Megaverse points in general. The 'orthogonal' circles around the point differentiate between the two types of neighborhoods.

The surface tension[22] of a universe γ satisfies the relation

$$\gamma = W/\Delta A \qquad (3.1)$$

where γ is expressed in erg/cm^2, W is the Work, and ΔA is the Area upon which the work is exerted. The pressure Δp exerted by the surface tension for a 'spherical' surface area is

---

[20] See Landau (1987).

[21] Although in actuality it is the universe that holds itself together by gravitation. The surface tension force is the result of the internal gravitation of mass-energy within the universe.

[22] A useful analogy: the Megaverse is a pool of water; a universe is a denser oil bubble within it. Surface tension caused by the cohesiveness of the oil molecules in the bubble makes it spherical (confines it to a spherical shape). Similarly a universe (denser than the Megaverse) is 'confined' within the Megaverse. This "classical" discussion differs from that presented in Blaha (2019c) which uses Hawking's definition of temperature.

$$\Delta p = 2\gamma/R \tag{3.2}$$

where R equals the radius of curvature of the surface. The above equations embody the concept that the surface tension force equals the pressure difference at the surface.

If the universe is flat then the surface pressure approaches ∞ giving confinement of the fields and particles to the universe:

$$R \to 0 \quad \text{implies } \Delta p \to \infty \tag{3.3}$$

Thus we have the theorem:

Theorem: A universe has no leakage of fields or particles into a higher dimensional space if the universe is exactly flat.

This theorem is particularly interesting in the case of our universe. It appears to be flat (or very close to flat). The flatness of our universe may be the reason no leakage of fields or particles from our universe has been detected to high accuracy.

If a universe is found with a non-zero radius of curvature[23] then one can expect that some fields and particles may emerge from it into the Megaverse.

While a zero radius of curvature prevents the exit of fields and particles from a universe, it does not prevent the entry of mass-energy into the universe from the Megaverse. Thus our Continuous Creation Model of chapter 14 in Blaha (2017c) may be relevant. Entry is possible; exit is forbidden in this case.

Eqs. 3.1 – 3.2 are reminiscent of the 'four laws for Black Holes' which are stated later.

3.4 Quantum Fields 'Emanating' from a Universe

If the curvature of the universe is zero (open universe) then no fields emanate from it. If the curvature of the universe is non-zero (closed universe) then fields may

---

[23] Points near Black Holes, neutron stars, and other large masses in our universe may have large curvature and may be points of leakage into the Megaverse – an interesting experimental question.

'leak' into the Megaverse. Then continuity conditions between a universe field and its Megaverse counterpart becomes of interest. We discuss this in detail later.

3.5 Universe Confinement by Conservation Laws

Every point in our universe is "infinitely" close to points of the Megaverse.

A universe occupies a region within the Megaverse. However because it is a lower dimension surface within the Megaverse the neighborhood of every point within a universe has an infinite number of Megaverse points that are not within the universe.

One might think that particles within a universe could then 'slip' into the Megaverse outside the universe with ease. However that is not the case. The laws of momentum conservation compel particles and interactions within a universe to be confined to the universe. More importantly, the Megaverse surface tension of a flat universe confines particles and fields within a universe.[24]

The only possible ways that a particle could exit from a universe are 1) if the particle collides with a particle with a momentum, some of whose components are in Megaverse dimensions extraneous to the universe's dimensions; or 2) a particle within the universe experiences forces with components in Megaverse dimensions extraneous to the universe. The first possibility exists if the Megaverse has a very low matter density outside of universes. The fact that this phenomenon has not been observed implies the Megaverse matter density is extremely low.

Thus conservation of momentum for particles and interactions, and surface tension, effectively confines particles within a universe even though the neighborhood of every point of a particle's trajectory contains an infinity of Megaverse points exterior to the universe. Similarly every interaction within a universe is confined when expanded in a Fourier series (assuming free fields) in universe coordinates.

---

[24] A useful analogy: the Megaverse is a pool of water; a universe is a denser, oil bubble within it. Surface tension caused by the cohesiveness of the oil molecules in the bubble makes it spherical (confines it to a spherical shape).

It is also possible for a Megaverse particle to enter[25] a universe through perhaps a collision that results in the particle being within the universe with momentum also solely within the universe. Thus the 'point boundary' of universes is porous. Particles can enter/exit a universe under appropriate conditions.

We conclude the 'point boundary' of a universe is not a barrier although surface tension force controls the entry/exit of particles and fields.

3.6 Objects Straddling a Universe-Megaverse Boundary

When a starship, or some other extended object, is entering/exiting a universe at some velocity the question of the state of the object arises. It is partially in and partially out of the universe. We know that the object being 4-dimensional will continue to be 4-dimensional, barring effects of forces that might "twist" parts of the object into additional dimensions.

There are also the more subtle quantum effects on the object due to the possibility of different quantization of the particles in a universe and the Megaverse. Quantization in different coordinate systems might result in different physical interpretations of matter. We resolved this issue by our PseudoQuantization Method described in Blaha (2018e). In its Appendix A we show that one can quantize using a form of Pseudoquantization that preserves (unitarily equivalent) particle interpretations in a universe and the Megaverse.

Thus extended objects can be partly in a universe, and partly in the external Megaverse without issues.

---

[25] In chapter 14 of Blaha (2017c) we consider a model that supplies a mechanism for the Hoyle-Narlikar continuous creation theory for the expansion of our universe. This model 'creates' mass-energy as an inflow from the Megaverse.

4. General Properties of the Megaverse

If one wishes to have a portrait of the Megaverse it seems likely that the it would contain universes that were scattered within it in a fashion similar to the scattering of galaxies within our universe—although on a much larger distance scale in multiple dimensions. In this chapter we will overview properties of the Megaverse and its universes. Much of this material previously appeared in Blaha (2017c) and in earlier Physics and starship travel books.

4.1 Megaverse Size, Lifetime, and Universe Separation

The first questions that naturally arise are the age, size, and general characteristics of universes within the Megaverse. In the absence of any experimental detail we will assume the relative size of entities (universes) in the Megaverse[26] is proportionate to the relative size of the entities (galaxies) in the universe.

(Average Galaxy Size)/(Universe Size) = (Average Universe Size)/(Megaverse Size)   (4.1)

Taking the average diameter of galaxies to be 400,000 light years, the age of the universe to be 13,800,000,000 light years, and the diameter of the universe to be 91.4 billion light years[27] (the estimated diameter of last scattering surface) we find the diameter of the Megaverse to be

$$\text{Diameter}_{\text{Megaverse}} = 2 \times 10^{16} \text{ light years} \qquad (4.2)$$
$$= 228,500 \times \text{Diameter}_{\text{Universe}}$$

---

[26] We assume the Megaverse is homogeneous at large distance scales. We also assume a proportionality between the average entity size (of galaxies in the case of unbiverses; and of umiverses in the case of the Megaverse) in part based on the relatively long lifetimes of universes and the Megaverse.

[27] There are much larger estimates of the universe's diameter based on the Inflation theory of A. Guth and others.

Using the 228,500 scale factor, we find the Megaverse age since the Megaverse purported 'Big Bang' to be

$$\text{Age}_{\text{Megaverse}} = 3\times`10^{15} \text{ years} \tag{4.3}$$
$$= 3 \text{ million billion years}$$

If the average separation between galaxies in our universe is 3,000,000 light years, then assuming distance scaling by 228,500, the average separation between universes in the Megaverse would be

$$\text{Separation}_{\text{Universes}} = 228{,}500 \times 3{,}000{,}000 \text{ ly} = 7 \times 10^{11} \text{ light years} \tag{4.4}$$
$$= 700 \text{ billion light years}$$

If we now assume the mass of a universe equals the total mass-energy of our universe[28] (including Dark mass and Dark energy) which is estimated to be $m_{\text{universe}} = 3 \times 10^{54}$ kg then the gravitational potential energy between two such universes separated by 700 billion light years is

$$V = G\, m_{\text{universe}}^2/\text{Separation}_{\text{Universes}} \tag{4.5}$$
$$= 9\times10^{70} \text{ kgm}^2\text{s}^{-2}$$

Then the gravitational force is

$$F = G\, m_{\text{universe}}^2/\text{Separation}_{\text{Universes}}^2 \tag{4.6}$$
$$= 1.35\times10^{43} \text{ kgms}^{-2}$$

and the resulting gravitational acceleration of universes is

$$a = 4.5\times10^{-12} \text{ m/s}^2 \tag{4.7}$$

The Baryonic and Leptonic forces associated with the Generation group of the Unified SuperStandard Theory may slightly modify the force between the universes. We view the small acceleration between universes due to gravity as physically acceptable. In a

---

[28] We assume our universe is 'average' in mass and size for the sake of discussion.

billion years the universe velocity would be $v = 1.4 \times 0^5$ m/s with $v/c = 0.0005$ and the distance traveled equal to 2.2×10^{21} m = 236,540 light years – negligible compared to the Separation$_{Universes}$. Universes would, on the average, only make contact after extraordinary long times. Thus we have a preliminary coherent view of Megaverse size and distance parameters.

4.2 Likely Features of the Megaverse

There are a number of features of the Megaverse that appear to be true.

4.2.1 Megaverse Curvature

The Megaverse has gravitation. Gravitation appears to be weak in the Megaverse so it is close to a flat space. The sources of Megaverse gravitation are the mass-energy of the universes within it and the density of mass-energy of particles outside of universes. Universes have a larger relative force of gravity due to a higher mass-energy density. Universes therefore have more curvature.

4.2.2 Megaverse Time Dimension

We will assume that the Megaverse has one complex time dimension denoted y^D for the simple reason that the absence of a time dimension would make the Megaverse static.

4.2.3 Megaverse Forces

In addition to Megaverse gravitation, the Megaverse has the forces in the Unified SuperStandard Model. These forces satisfy continuity conditions at universe boundaries.

4.2.4 Megaverse Parameters

Megaverse physical constants and particle masses have the same values as in our universe due to continuity.

4.2.5 Megaverse Vacuum Fluctuations

Megaverse Vacuum fluctuations may be a source of the generation of universes and particles. Vacuum fluctuations might account for the Big Bang. The time scale for the persistence of universes generated by a vacuum fluctuation is likely to be an extrapolation of vacuum fluctuation persistence within our universe.

4.2.6 Megaverse Matter and Chemistry

The existence of many more dimensions in the Megaverse suggests that multi-dimensional forms of matter and energy could exist *between* universes. As a result such Megaverse atoms, compounds and Chemistry will be very different and much more varied than in our universe. If such matter exists in the Megaverse then 'mining' such matter for use in our universe—would give us exotic new compounds and Chemistry that would be partially inside and partially outside, of our universe.

This possibility makes venturing into the Megaverse economically and scientifically desirable since such materials cannot be created within our universe.

4.3 Features of Universes within the Megaverse

We know of our universe from the 'inside.' However the features of our universe from a Megaverse perspective are not at all certain. In this section we will describe the Megaverse view of a universe's properties.

We shall assume a universe is a closed or open surface within the Megaverse of much higher mass-energy density than the Megaverse's mass-energy density by perhaps as much as a factor of 10^{30}. Chapter 14 of Blaha (2017c) describes a model of Megaverse mass-energy inflow into our universe exemplifying this feature.

4.3.1 Universe Area and Mass

The mass of a universe is an important property since mass is one of the sources of Megaverse gravitation, and interaction between universes.

While a universe is not believed to be a black hole (although Hawking has recently jokingly? suggested that our universe may be a black hole, and even more recently suggested black holes are not quite black holes – grey?), there are general qualitative similarities that lead us to consider the possibility that the four laws of black holes[29] may apply in part (or their entirety) to universes. In particular the 2nd law states

$$dM = \kappa dA/8\pi + \Omega dJ \qquad (4.8)$$

where dM is the change in "mass/energy," A is the area of the Black Hole (universe), Ω is its angular velocity and J is the angular momentum.[30] From eq. 4.8 it appears we can reasonably define a "mass" for a universe in terms of a universe's area:

$$M = \kappa A/8\pi \qquad (4.9)$$

This definition seems to capture the physics of universes that could be used in developing a dynamics of universes as we do later. It allows us to escape the dilemma of having zero total energy for universes that would preclude treating universes as particles in the Megaverse and developing a Megaverse dynamics of universe-particles. Later we will also show how to define a mass for a universe that is time dependent.

4.3.2 Relation between Universe and Megaverse Vector Fields

In Blaha (2018e) we defined the interactions of The Unified SuperStandard Theory. These interactions and their groups also exist in the Megaverse. The Fourier expansions of fields in the Megaverse are different. They must be expressed in terms of the D coordinates of the Megaverse.

Within a universe, since each point of the universe is surrounded by Megaverse points, a field has both an expression in universe coordinates, x, and an expression in Megaverse coordinates, denoted y. For a vector field in the universe $A_U^{\mu}(x)$ there is an equivalent representation of the field in Megaverse coordinates $A_M^i(y)$. These representations are related by a coordinate transformation. If we define a map from universe coordinates to Megaverse coordinates with

[29] Wald, R. M., "The Thermodynamics of Black Holes", *Living Reviews in Relativity* **4** (6): 12119 (2001).

[30] Although the angular momentum of a universe is not measurable if there is only one universe (as DeWitt argued in a quote earlier), the existence of multiple universes within the Megaverse enables the relative angular momentum of a universe to be determined.

$$y^i = f^i(x) \qquad (4.10)$$

then the field representations are related by[31]

$$A_M{}^i(y) = \partial y^i/\partial x^\mu\, A_U{}^\mu(x) \qquad (4.11)$$
$$= \partial y^i/\partial x^\mu\, A_U{}^\mu(f^{-1}(y))$$

in the domain of the universe. The values of the field at Megaverse points in a neighborhood of a universe point are determined by continuity.

Outside the domain of a universe the Megaverse field value is determined by its sources.

At the boundary of a universe there must be continuity in the expectation values of the Megaverse fields.

4.3.3 Megaverse Gravitation and Free Matter

We assume that a generalization of Einstein's theory of Gravity exists in the Megaverse.

Just as our universe has matter and radiation between galaxies, it seems reasonable to assume that 'free' matter and radiation exists in the Megaverse outside of universes. Such mass-energy would have two roles: to gravitationally affect the dynamics of the Megaverse and the motion of universes within it, and to possibly fuel the expansion of universes. The expansion of our universe may be due to an influx of matter and energy from the external Megaverse. Many years ago Hoyle and Narlikar considered the possibility of 'continuous creation of matter.' We suggest that an influx of Megaverse matter may be the actual source. We considered this possibility in chapter 14 of Blaha (2017c), which contains an unpublished paper by this author written approximately seven years previously.

---

[31] Implicit in eq. 4.10 is an inverse relation $x^\mu = f^{-1}(y)$ which is necessarily based on a restriction of the y-coordinates to obtain a 1:1 relation between the y and x coordinates. The restriction is best implemented by requiring the domain of y coordinates be restricted to those y coordinates within the universe surface. The result is a 1:1 relation between the y-domain coordinates and the x universe coordinates.

Thus we arrive at a view of the Megaverse of matter and universes that is analogous to our universe of galaxies.

4.3.4 Expansion of Universes

Our universe expanded from the Big Bang to its current size and is still expanding. It is likely that other universes have undergone similar expansions. According to chapter 32 of Blaha (2018e) there is an infinite surface tension at the boundary of our universe. How can the universe have expanded, and continue expanding, under such conditions. We see a two phase expansion of the universe.

For a period of time after the Big Bang the universe did not have an infinite surface tension, thus hindering universe expansion into the Megaverse due to surface tension force. The cause is an effect discovered by Eŏtvos – a temperature dependence of the surface tension force. Eŏtvos pointed out a critical temperature T_c existed that caused the surface tension force to decline as the temperature increased:

$$\gamma V^{2/3} = k(T_c - T) \tag{4.12}$$

where k is the Eŏtvos constant, and V is the volume of the universe (the liquid 'drop') Assuming a spherical universe, the volume is

$$V = 4\pi R^3/3. \tag{4.13}$$

Thus

$$\gamma = (4\pi/3)^{-2/3} k(T_c - T)R^{-2} \tag{4.14}$$

For very high temperatures such as existed after the Big Bang $T > T_c$ γ would be negative indicating that an outward pressure existed from the universe into the Megaverse promoting expansion. Thus in the high temperature period $(T > T_c)$ after the Big Bang the surface tension force favored expansion of the universe.

After this phase, the surface tension γ is positive. The Megaverse is then superficially partially 'impeding' expansion due to surface tension. However, the surface tension pressure of the Megaverse causes leakage *into* the universe from the

Megaverse causing its mass to increase, and its radius and volume to increase. Expansion due to the accretion of Megaverse mass-energy!

The above scenario is supported by the two phase model suggested by section 14.15 consisting of a Big Bang expansion model (chapters 11 – 13), and a mass-energy accretion model (chapter 14) – all of Blaha (2017c). Note as the radius of curvature goes to zero, eq. 4.14 suggests an increasing surface tension pressure 'pushing' particles into the universe.

Thus a complete universe expansion scenario is evident based on surface tension physics and the Big Bang.

4.3.5 Universe Generation from Vacuum Fluctuations

Vacuum fluctuations could generate universe-antiuniverse pairs. Antiuniverses would have certain 'negative quantum numbers.' We view universes/antiuniverses, when created (Big Bangs), as 'ultra-small' particles of large mass-energy that can proceed to expand to great size. If they are generated by a vacuum fluctuation they will, after possibly a certain time, recombine into the vacuum.

4.3.6 Universes as Black Holes

It is conceivable that a universe could be so dense and confined that it would be effectively a Black Hole. In this case the internal quantum numbers of the Black Hole universe would be inaccessible and only it's mass, velocity and angular momentum would be observables.

4.3.7 Life in Other Universes

It is likely that life exists in some if not all of the universes of the Megaverse. If the physical constants, laws, and masses of the interior of a universe are similar if not identical to ours then the possibility of intelligent species, even human-like species, is very likely. This possibility is an important motivation for humanity to reach for the Megaverse.

5. On Exiting From Our Universe

The first question on realizing that one is in a confined space usually is "How do I get out?" This question takes on a dramatic new aspect when we ask "How do we exit from our universe?" Like tadpoles in a pool of water it is a question of importance since universes, like pools of water, may 'evaporate' – a possibility raised from time to time by cosmologists.

In this chapter we present a mechanism for exiting our universe realizing that it could not be developed and used for perhaps *tens of thousands of years*.[32]

5.1 Looking for an Exit Point

Due to our discussion of surface tension at the boundary of the universe the first issue that we will address is whether there are 'points' where egress from our universe is facilitated. We will see that the vicinity of bodies with large gravitational fields such as neutron stars and Black Holes seriously weaken the nearby universe surface tension and are thus the best locations for exiting our universe into the Megaverse.

5.2 Neutron Star (Black Hole) Exit Point Surface Tension

Near a neutron star, or similar small, very massive body, the curvature of the universe (which is close to zero – flat) becomes significant. Then, as we saw in chapter 32, of Blaha (2018e) the surface tension force becomes possibly much smaller (not infinite!) and the force keeping mass-energy within the universe decreases enabling a starship to conveniently exit into the Megaverse.

---

[32] This time frame could be shortened if it were possible to use faster-than-light starships to execute the maneuvers described in Blaha (2018e) around a Black Hole (or neutron star) without the destruction of the starship and its passengers. If faster than light starships become a reality then exits from the universe may become feasible only thousands of years afterwards.

In chapter 32 (Blaha (2018e)) we saw the pressure exerted by the surface tension for a 'spherical' surface area is

$$\Delta p = 2\gamma/R \qquad (5.1)$$

where R equals the radius of curvature of the surface with R = 0 for a flat space. A non-zero value for R causes the infinite force of flat space-time to become a finite force which the starship can overcome to exit.

Later in that chapter we described a slingshot maneuver for a starship in which a starship 'whips' around a neutron star using its gravity to twist into the Megaverse. The trajectory is depicted in its Fig. 5.1a. We call starships with this capability *uniships* since their ultimate goal is to travel to other universes.[33]

A neutron star (Black Hole) slingshot has an additional advantage for transit to the Megaverse. This advantage is enhanced by an ultra-fast uniship speed. An extremely large speed causes the uniship to appear 'hot' using the conversion between velocity and temperature. A high speed will create a 'hotspot' in the universe boundary surface that could substantially lower the surface tension force.[34] Eőtvos showed that the surface tension force has a critical temperature T_c which satisfies:

$$\gamma V^{2/3} = k(T_c - T) \qquad (5.2)$$

where k is the Eőtvos constant and V is the volume of the universe (the liquid 'drop'). As the uniship speed increases to an equivalent temperature beyond T_c, the Megaverse surface tension force will drop to zero, and the uniship can freely enter the Megaverse.

---

[33] The design and human factors of uniships: living conditions, travel times, communications from the Megverse, long-lived ship materials, suspended animation, and so on are described in Blaha (2017c) and earlier books.

[34] Much more importantly is the temperature in the region around a neutron star. An idea of the temperatures involved is the magnitude of the temperature in a region 1 – 7 km in size in a colliding neutron stars event. The temperature was estimated to be the extraordinary temperature 350 GK (gigakelvins) by R. Oechslin et al, arXiv:astro-ph/0507099 (2006). Temperatures of the order of 10^{12} K would seem to severely lessen or eliminate surface tension.

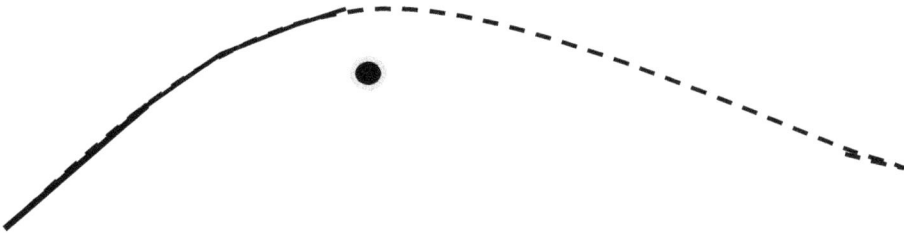

Figure 5.1a. The trajectory of a uniship in a slingshot maneuver with a neutron star or Black Hole (the dark circle). The repulsive baryonic force causes the "turn" away from the star/Black Hole into the Megaverse. The solid line corresponds to the time in which the uniship is wholly within the universe and dominated by gravity. The dotted line reflects the transition of the uniship into the Megaverse.

Therefore the Megaverse – universe interface can be viewed as a local surface bubbling that thins out at high speed to nothing – no barrier force.

We can estimate the critical velocity v_c as a function of the critical temperature, at which the surface tension force goes to zero, using the velocity maximum of the superluminal Maxwell-Boltzmann distribution:[35]

$$v_c \approx c + \tfrac{1}{2}\, m^2 c^5 / (2kT_c)^2 \qquad (5.3)$$

where m is the mass of the uniship (presumably hundreds of thousands of metric tons). The critical temperature can also be expected to be extremely high. As a result the ratio, the critical velocity, is likely to be large. At sufficiently high superluminal velocity there is no barrier to entry to the Megaverse.

The precise determination of the critical temperature remains to be done. Chapter 14 of Blaha (2017c) estimates the critical temperature for our models in Blaha (2018e).

---

[35] Blaha (2017c) section 1.4.3.

SECTION 2: QUANTUM BIG_BANG SECTOR

This section presents a Quantum theory of the Big Bang combined with a Megaverse Driven Expansion due to an influx of mass-energy from the Megaverse.

6. Universe Particle Dynamics

The origin of our universe and other universes is a weighty question that has been considered in a number of theories. Our view is that the existence of the Megaverse, and the locality of universes (in the one case of which we are certain), suggest the Megaverse is the "platform" of a type of particle physics in which universes play the role of particles.[36] There are clear points of difference: particles have a fixed mass, particles have well-defined quantum numbers, and so on. Nevertheless we can consider universes to be massive particles, *universe particles*, with a dynamics that is primarily based on gravitation.

This chapter describes aspects of the interactions of universe particles due to gravitation, baryon number forces, and collisions between universes. It also the describes the genesis of universes due to vacuum fluctuations, the fission of universes, and the internal distortion of universes due to acceleration and the presence of 'nearby' universes. Section 4 examines universe particle dynamics in detail including vacuum polarization of universe particles.

6.1 The Internal Distortion of Universes

In the absence of external forces universes are considered to be uniform in the large. However the acceleration of a universe in the Megaverse can distort the universe. Also the existence of a nearby universe(s) could cause the uniformity of a universe to be lost due to gravitation and baryonic forces.[37]

---

[36] This view will be significantly supported by the findings of section 4.
[37] The baryonic forces, the Baryonic force and the Dark Baryonic force, on a universe are large due to their additivity in our universe and other universes.

6.1.1 Impact of Universe Particle Acceleration – Lopsided Internal Structure of Universe

Universes can accelerate within the Megaverse due to external Megaverse forces. Universe acceleration should be detectable within a universe as "lopsidedness." There would be a shift of parts of the universe opposite to the direction of acceleration resulting in a difference in the features of the universe "in front" compared to those "in back" – an acceleration effect just as one sees when a jet accelerates.

Interestingly recent data from the Planck observatory of the European Space Agency confirms and extends earlier data from NASA's WMAP observatory that one side of the universe appears different from the other side. There are temperature differences and mass distribution differences – just as one might expect if the universe were accelerating as a unit.

Thus we see the beginning of data suggesting our universe may be accelerating through the Megaverse. Some Planck observatory scientists have suggested their data is a preliminary indication of the Megaverse.

6.1.2 Impact of External Forces on Universe Structure

The presence of a nearby universe could cause a universe to lose its large scale uniformity and the mass-energy of the universe to drift over time to the 'nearby' side of the universe due to gravitation and baryonic forces.

6.2 Universes in Collision

We can assume that the dynamics of universes in collision will be analogous to that of galaxies in collision since gravity is a dominant force in both cases. Colliding galaxies have often been observed. Their dynamics should provide guidance for the case of universes in collision.[38]

It is clear in the case of colliding galaxies, and of colliding large nuclei (gold and lead typically) that there are several types of collisions with differing results. Similarly, the types of universe collisions can be qualitatively classified as:

---

[38] The high energy collision of atomic nuclei at Brookhaven, CERN and other laboratories also is analogous in overall detail with universes in collision.

1. Clean collisions in which universes nudge each other but retain their identity. These are extreme peripheral collisions. If the universes overlap slightly then the typically spherical symmetry of the universes may become distorted and they may become lopsided.[39]

2. Peripheral collisions in which the universes retain their identity but are connected by a trailing string of mass-energy. Eventually the string breaks and the universes separate. Subsequently the pieces of trailing string in each universe contract due to their universe's gravitational effects and perhaps form new "bubble" universes.

3. Two universes can collide and produce multiple universes.

4. Two universes can collide in a "central" collision and amalgamate into one universe. They can intermix with both the baryonic, gauge, and gravitational forces causing a redistribution of their masses. They may separate afterwards or may coalesce into a single universe. One result of this may be lopsided universes. Our universe appears to be lopsided. Some cosmologists believe this is due to a near collision of our universe with another shortly after the Big Bang.

6.3 Creation of Universes through Gauge Field Fluctuations

One of the most exciting questions in Cosmology is the origin of our universe. The conventional view is that it originated in a Big Bang from an infinitesimal point in space. The source of the Big Bang and the prior state of the Megaverse, if there was one, is the subject of much speculation. Based on the particle interpretation of the Wheeler-DeWitt equation, the possibility of a baryonic force strongly supported by conservation of baryon number, and the Megaverse concept, it is reasonable to consider the possibility that the universe originated in a vacuum fluctuation.

---

[39] The Wilkinson Microwave Anisotropy Probe (WMAP) and the Planck European Space Agency satellite have been accumulating data since 2001 that suggests the universe may be lopsided with hot and cold spots on opposite sides of the universe differing from those on the other side being hotter and colder respectively—*perhaps the result of a collision when the universe was young.*

In this case there would be two Big Bangs one for our universe and one for an anti-universe. One would expect that they would have opposite corresponding features: one with baryon dominance – one with anti-baryon dominance, and one left-handed – one right-handed.

Our formulation of universe particle theory allows for the generation of a universe particle and anti-particle as a vacuum fluctuation. We view a universe particle as having a substantial excess of baryons, N, as we see in our universe. Its anti-universe at the time of creation (the Big Bang point) is its "mirror image" having the "same" number of anti-baryons (baryon number –N) so that baryon number is conserved by the fluctuation event. Thus the excesses of one universe are compensated by the excesses of the other.

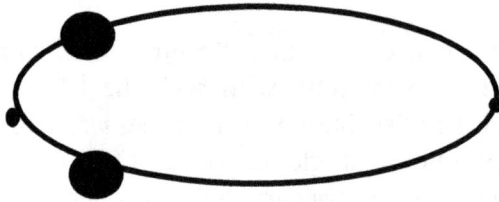

Figure 6.1. Generation of a universe – anti-universe pair as a vacuum fluctuation.

The small value of the coupling constant should lead to an extremely long lifetime for the universes generated by the fluctuation. Thus the 45 billion year life of our universe is not unreasonable. The probability of the creation of universes by vacuum fluctuations should be correspondingly small.

The sizes of the created universe and anti-universe should be very small just as Big Bang theories of our universe suppose.

6.4 Fission of Universes

Under certain circumstances the distribution of matter in the universe may lead to the fission of the universe into two separate universes. Our theory supports this possibility for universe particles. The detailed mechanism of the fission process is not specified by the model.

6.4.1 Fission of Normal Universes

The fission of universe particles in our universe particle model is depicted in the Feynman diagram in Fig. 6.2.

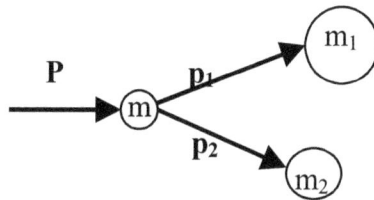

Figure 6.2. Fission of a universe particle into two universe particles.

The sum of the masses of the output universe particles is usually less than the original universe particle mass. However if the fission takes a long time and the masses are time dependent then the produced universe particles combined masses may exceed the original universe's mass.

6.4.2 Tachyon Universe Particle Fission to More Massive Universe Particles

In Blaha (2007a) we showed that a tachyonic (faster than light) particle could fission into particles of larger mass through the conversion of momentum into mass. In this section we show that a tachyonic universe particle may fission into two more massive universe particles.[40] This phenomenon is of particular interest because it enables tachyonic universes to spawn in a new novel way not previously considered in discussions of the origin of universes.

A simple model lagrangian[41] for a tachyonic universe particle is

---

[40] We will use the term mass here to denote mass-energy. Since we identified mass as a multiple of area earlier, the comments here would appear to apply to universe area as well.

[41] See Blaha (2018e) and earlier books by the author for a detailed discussion.

$$\mathcal{L}_{\parallel} = \psi_T^S(Y(y))[\gamma^\mu \partial/\partial y^\mu - e_B \gamma^\mu B_{u\mu}(Y(y)) - m(t)]\psi_T(Y(y)) - \tfrac{1}{4} F_{Bu}^{\;\mu\nu}(Y(y))F_{Bu\mu\nu}(Y(y))$$
$$- \tfrac{1}{4} F_u^{\;\mu\nu}(Y(y))F_{u\mu\nu}(Y(y))$$

We assume m(t) is constant.

When a particle or a universe particle fissions (decays) one normally expects that the masses of the particles or universe particles produced by the decay to be smaller than the mass of the original particle or nucleus. In the case of tachyonic (faster-than-light) elementary particles, or universe particles, a much different possibility is present: a tachyon universe can decay into heavier tachyons (perhaps through a distortion of the universe internally into two 'lumps'.) We consider the specific case of a tachyon universe particle decaying into two universe particles whose total mass is greater than the original. (See Fig. 6.3.)

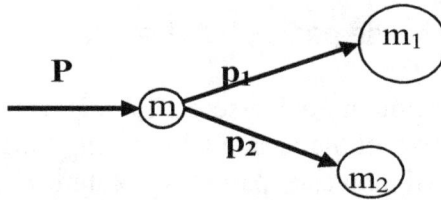

Figure 6.3. Two universe particle decay of a tachyon universe particle.

We will assume the initial tachyon universe particle has zero energy ($p^D = 0$) and thus the tachyons universe particles emerging from the decay also have total universe particle energy zero. The analysis is based on conservation of total universe energy and momentum in Megaverse space outside of universes. The below discussion applies to D-dimensional space with (D – 1)-dimensional spatial coordinates.

Momentum conservation implies

$$\mathbf{P} = \mathbf{p}_1 + \mathbf{p}_2$$

Since all energies are zero

$$(c\mathbf{P})^2 = (c\mathbf{P})^2 = m^2$$
$$(c\mathbf{p}_1)^2 = (c\mathbf{p}_1)^2 = m_1^2$$

$$(cp_2)^2 = (c\mathbf{p}_2)^2 = m_2^2$$

where $P = |\mathbf{P}|$, $p_1 = |\mathbf{p}_1|$, and $p_2 = |\mathbf{p}_2|$. If we now square the above equation for $\mathbf{P}$ and then use the above three equations we obtain

$$m^2 = m_1^2 + m_2^2 + 2m_1m_2 \cos\theta$$

where θ is the opening angle between the emerging universe particles momenta of $\mathbf{p}_1$ and $\mathbf{p}_2$. There are a number of interesting cases:

Case $\theta = 0$:

$$m = m_1 + m_2$$

The masses of the outgoing universe particles sum to the mass of the original tachyon universe particle.

Case $\theta = \pi/2$:

$$m^2 = m_1^2 + m_2^2$$

The masses of each outgoing universe particle tachyon are less than the mass of the original tachyon universe particle.

Case $\theta = \pi$:

$$m^2 = (m_1 - m_2)^2$$

In this case either $m_1 > m$ or $m_2 > m$. Thus one of the outgoing tachyon universe particles has a greater mass than the original tachyon universe particle. Mass is effectively created from the spatial momentum of the initial universe particle. This process is the inverse of normal particle and universe particle fission where the sum of the outgoing masses is always less than the original particle's mass and the difference is mass converted into energy in the form of additional photons.

This last case, where one of the outgoing universe particles is more massive than the original universe particle, is not just for $\theta = \pi$. Since

$$\cos \theta = (m^2 - m_1^2 - m_2^2)/(2m_1m_2)$$

we see that the sum of the outgoing universe particle masses is always greater than the original tachyon universe particle *mass (except when $\theta = 0$)* since

$$\cos \theta = 1 + [m^2 - (m_1 + m_2)^2]/(2m_1m_2) \leq 1$$

and thus

$$[m^2 - (m_1 + m_2)^2]/(2m_1m_2) \leq 0$$

Note $m = m_1 + m_2$ only if $\theta = 0$.

Since we can transform the above discussion to the case of universe particle tachyons having non-zero Megaverse energy using an ordinary D-dimensional Lorentz transformation, the discussion in this subsection is general.

We therefore conclude that when a tachyon universe particle decays into two tachyon universe particles the sum of the masses of the produced tachyon universe particles is greater than the mass of the original tachyon universe particle except if the angle between the momenta of the produced tachyon universe particles is zero. In that case the sum of the masses of the produced tachyon equals the mass of the original tachyon universe particle and the produced universe particles overlap.

7. A Driving Force for Universe Expansion

Some years ago the author introduced a new method[42] that ensured perturbation theory calculations were always finite to all orders in perturbation theory. The method was based on a generalization of space-time coordinates to have imaginary q-number parts that consisted of a massless vector quantum field theory that was similar in form to the quantum electromagnetic field. We called these coordinates Two-Tier coordinates because they were based on an extended two level extremum principle.

We showed that these quantum smeared coordinates of Two-Tier Quantum Field Theory succeeded in removing all ultra-violet infinities in perturbation theory including the fermion triangle infinities. Remarkably the high precision, low energy[43] predictions of QED remained true in Two-Tier QED and thus remained consistent with experiment to a hitherto unsurpassed level of accuracy. 'Low' energy predictions in other quantum field theories also remained unchanged. At high energies, Two-Tier perturbation theory results are finite and consequently all ultra-violet infinities, to any order in perturbation theory, in *any number of space-time dimensions* were eliminated.

7.1 Additional Benefits of Two-Tier Coordinates

In addition to removing perturbation theory infinities Two-Tier coordinates have the features:

1. They enable us to define   finite theories of Quantum Gravity and 'non-renormalizable' quantum field theories based on polynomial lagrangians.

2. They enable us to tame vacuum fluctuations.

---

[42] See Blaha (2002) and (2005a).
[43] Relative to a mass scale that was perhaps of the order of the Planck mass.

3. They enable us to eliminate infinities associated with the Big Bang.

4. They provide a mechanism to generate the explosive growth of the universe in a role as part of Dark Energy.[44]

Items 3 and 4 were proved in earlier books.[45] They will be part of a new extended theory of universes in this book. We will see Two-Tier Quantum Field Theory is required at the most fundamental level.

7.2 Two-Tier Features in 4-Dimensional Space-Time

Two-Tier Quantum Field Theory,[46] is based on a new method in the Calculus of Variations. It uses two 'layers' of fields to introduce quantum coordinates. We shall consider this technique briefly using a massless vector field $Y^{\mu}(y)$ analogous to the electromagnetic field.

In 4-dimensional space-time the massless vector field has the form $Y^{\mu}(y)$ where the index μ ranges from 0 through 3. The X^{μ} coordinate system, where it appears, has a c-number real part and a q-number imaginary part. Thus particle fields which are normally defined on four-dimensional real space-time are now defined on a complex four-dimensional space-time where four imaginary dimensions will appear as *Quantum Dimensions* embodied in a vector quantum field $Y^{\mu}(y)$:

$$X^{\mu}(y) = y^{\mu} + i\, Y^{\mu}(y)/M_c^{\,2} \qquad (7.1)$$

where M_c is an extremely large mass of the order of the Planck mass or perhaps much larger.

---

[44] See Blaha (2017b) and earlier books for details. This section is basically a summary of some features.
[45] See Blaha (2004): *Quantum Big Bang Cosmology.*
[46] See Blaha (2005a), and Blaha (2002), for discussions of this new method to eliminate infinities in quantum field theory calculations.

The $Y^{\mu}(y)$ field is a function of the subspace y coordinates. The real part of the space-time dimensions will be taken to be the space of real-valued y coordinates.[47]

The imaginary part of space-time coordinates is the massless $Y^{\mu}(y)$ vector quantum field that is suppressed further by a very large mass scale that reduces the imaginary Quantum Dimensions to the infinitesimal except at large momenta. The effects of Quantum Dimensions only become appreciable in quantum field theory at energies of the order of M_c. At these energies exponential Gaussian factors in each particle (and ghost) propagator are generated by the Quantum Dimensions and serve to make perturbation theory calculations ultra-violet finite – including calculations in Quantum Gravity.

The formalism introduces a new form of interaction that does not have the form of the simple polynomial interactions that have hitherto dominated quantum field theories. This form of interaction takes place via the composition of quantum fields and can be called a *Dimensional Interaction* or an *Interdimensional Interaction* since it affects particle behavior through Quantum Dimensions.

The basic ansatz of the Two-Tier formalism is to replace every appearance of a coordinate x in a quantum field with the variable

$$x^{\mu} \rightarrow X^{\mu} = (y^0, \mathbf{y} + \mathbf{Y}(y^0, \mathbf{y})/M_c^2) \tag{7.2}$$

where $\mathbf{Y}(y^0, \mathbf{y})$ is the spatial part of a free massless vector field with features that are identical to the free QED field in the Radiation gauge.

Then one finds that the momentum space free field Feynman propagators $G(k)$ of all particles acquires a Gaussian factor $\exp(h(k))$:

$$G(k) \rightarrow G(k)\exp(h(k)) \tag{7.3}$$

so that all perturbation theory diagrams are finite. The result is finite perturbative results for all calculations to any order in perturbation theory.

---

[47] In a deeper theory the real part might also be a quantum field that undergoes a condensation to generate c-number coordinates. We will not consider this possibility in this book.

Blaha (2005a) shows that Two-Tier theories are finite, Poincare covariant, and unitary. (See Blaha (2005a), chapter 5, for a complete discussion.)

7.3 Importance of Two-Tier Coordinates

Two-Tier coordinates can be used in an extension of the Robertson-Walker Model to create a Big_Bang sector that prevents infinities at the Big Bang point in time. The fuzziness of these coordinates effectively prevents a "condensation" of a universe to a point going backward in time from the present as many current Big Bang models do. See the following chapters for details. The role of Quantum Mechanics in preventing the collapse of hydrogen atoms to a "divergent" point is analogous.

8. Plan for this Section's Theoretic Presentation

The plan of the development of the theory begins with the Robertson-Walker Model in real space-time. We then proceed to generalize the Robertson-Walker Model to have q-number coordinates consisting of a real-valued part and an imaginary part that is a vector field (similar to free QED) that introduces a repulsive effect at extremely short distances. This effect prevents a collapse at the Big_Bang to a point with infinite mass-energy and temperature.

The motivation for this construct is analogous to the motivation that appeared when atomic models were being constructed at the beginning of Quantum Mechanics. Quantum Mechanics prevented the collapse of an atom to a point as circling classical electrons spiraled to the central nucleus. In the present case the universe is prevented from beginning from a point by the imaginary quantum field that provides a form of "anti-gravity" at ultra-short distances.

The introduction of an imaginary quantum part of coordinates is not a form of "Dark Energy" and yet it strongly influences the initial start and growth of the universe.

Having defined a form of quantum coordinates for the Robertson-Walker Model, we then proceed to calculate the growth of the universe from $t = 0$ using the known mass-energy with the proviso that we treat Dark Matter as similar to ordinary matter in its effect on universe growth (the Hubble constant).

We also introduce a new term in the total mass-energy for the effect of universe surface tension at the boundary of the Megaverse. In doing this, we note that every point of the universe is in a Megaverse neighborhood. Thus the surface tension energy contribution pervades the entire universe and causes an inflow of Megaverse mass-energy throughout the universe. As a result the universe expands more rapidly and the Hubble constant increases with time.

We calculate the Hubble Constant for all time starting as finite at $t = 0$ and growing in time as we presently observe. We also calculate the radius and temperature of the universe in the large as a function of time. Much of this theory appeared in Blaha (2004).

9. Known Mass-Energy Growth Factors for Universes

In chapter 6 we discussed some possible seeds from which universes might spring. One important possibility is as a vacuum fluctuation that produces a universe-anti-universe pair. (See section 4.) An attractive feature of this mechanism is that it accounts for the preponderance of masses in our universe as well as other features such as left-right asymmetry. There are other proposals for the origin of our universe that merit attention such as an oscillating universe scenario.

Whatever the source of our universe, it is clear that it began as a very small region (or a point) containing a large amount of mass-energy. So we began with that picture. We then developed a theory of universe evolution to its current enormous (and growing) size. The theory is based on a generalized Robertson-Walker Model described beginning in the next chapter that incorporates complex-valued coordinates.[48]

The propellants for the expansion of the universe are: 1) its mass-energy contents; 2) Quantum effects of Two-Tier coordinates; and 3) an influx of mass-energy into the universe from the Megaverse. Together these factors combine to give an initial universe metastable state of finite size, to generate growth as represented by the Hubble constant, and to cause the Hubble constant to increase with time (as appears experimentally). If any of these three growth elements are absent, it becomes more difficult to understand the observed growing size of the Hubble constant. Currently there are numerous speculations as to the origin of the Hubble constant growth. Most of these speculations are complex and somewhat hypothetical. This quantum model leads clearly to the observed universe and its growth.

Our relatively simple model resolves the mystery of Hubble constant growth.

---

[48] We describe aspects of Complex General Relativity in Blaha (2018e).

9.1 Quantum Effects of Complex Two-Tier Coordinates

Chapter 7 discussed the importance of Two-Tier quantum coordinates in affecting the growth of the universe—especially in its initial stages near t = 0, although they have an effect up to the present. Two-Tier quantum coordinates are a significant part of our model although they do not directly contribute to the mass-energy fueling universe expansion.

We will now describe the components of a universe's mass-energy driving expansion. We assume the universe is uniform in the large throughout its lifetime.

9.2 Universe Energy Density Components

The universe, as we know it today, contains a variety of forms of energy. The current densities of these forms of energy will be seen later. From them we can develop the form of the total energy density and project it back to the instant of the Big Bang. The Big Bang period is significantly different as we will see. The total energy density is composed of four major parts: radiation energy density, mass density, dark energy, and Megaverse mass-energy inflow, which all depend on time t.

9.2.1 "Electromagnetic" Energy Density

The universe's radiation energy density $\Omega_\Gamma = \rho_\Gamma/\rho_{crit}$ is[49]

$$\Omega_\Gamma(t) = \Omega_\gamma(t) \tag{9.1}$$

where ρ_Γ is the universe's electromagnetic energy density, and ρ_{crit} is the critical density of the universe.

9.2.2 Mass Density

The universe's mass density $\Omega_M = \rho_M/\rho_{crit}$ is also composed of two parts; the "visible" mass density Ω_m and the "Dark" mass density Ω_d, which makes its presence known indirectly in astrophysical observations. In specifying Ω_M we assume the Dark

---

[49] ρ_{crit} is the critical density of the universe. $\rho_{crit} = 1.05371(5) \times 10^{-5}$ h^2 (GeV/c^2) cm^{-3} where h =0.678(9) according to M. Tanabashi *et al* (Particle Data Group), Phys. Rev. D**98**, 030001 (2018).

and normal mass densities have the same time dependence and gravitational characteristics unlike some other models of universe expansion:

$$\Omega_M(t) = \Omega_m(t) + \Omega_d(t) \tag{9.2}$$

9.2.3 Dark Energy Density

The dark energy (Λ) density Ω_Λ = .692.

9.2.4 Additional? Energy Density: Surface Tension "Energy" Density, Etc. Ω_T

A possible? fourth major contribution to the total mass-energy density is 1) an inflow of mass-energy from the Megaverse to every point in the universe, and 2) an additional energy density due possibly to quintessence (section 4).

As we described in Blaha (2018e), and earlier books, each point in the universe is within a neighborhood of Megaverse points. Thus each point in the universe is on the universe's boundary with the Megaverse. There is an inherent inward directed surface tension at each point due to the gravitation of the mass-energy within the universe. The gravitational force is directed towards the "center" of the universe just as the earth's gravity is cumulatively directed towards the center of the earth.

The surface tension force causes Megaverse mass-energy to enter the universe in extremely minute amounts per unit volume. However the enormity of the universe creates a great continuing increase in the universe's mass-energy.[50]

The ingredients of the surface tension energy density inflow are:[51]

1. The gravitational (surface tension) force on the universe's surface is

---

[50] Hoyle and Narlikar proposed a Steady State model of the universe of this sort. However they could not account for the source of the influx since they were unaware of the Megaverse and the consequent surface tension effect. See H. Bondi and T. Gold, M.N.R.A.S. **108**, 252 (1948); F. Hoyle, M.N.R.A.S. **108**, 372 (1948); F. Hoyle and J. V. Narlikar, Proc. R. Soc. London **A290**, 162 (1966) and references therein.

[51] In this development we use a diferent formulation of universe surface tension, which seems more appropriate. The earlier Eötvos formulation is an alternate possibility. The conclusions reached later for the formulation here presented are similar to those that would be reached with the Eötvos formulation.

$$\kappa = \rho_M(t)V(t) = 2\pi^2 G\rho_{tot}(t)R(t) \qquad (9.3)$$

in the Robertson-Walker model where $\rho_M(t)$ is the universe average density.

2. The surface tension γ of the universe is[52]

$$\gamma = k_E T_H/(4\pi R^2)^{-1} \qquad (9.4)$$

where

$$T_H = \kappa/2\pi$$

where T_H is the Hawking black hole temperature, k_E is a constant, and R is the radius of curvature. In the Robertson-Walker model the radius is $R = k^{-\frac{1}{2}}a(t)$ where k is the Robertson-Walker constant.

Putting these concepts together we estimate the surface tension mass-energy density Ω_T = ρ_T/ρ_{crit} inflow per unit time at time t to be

$$\Omega_T(t) = \Omega_{TST} + \Omega_{Trest}$$
$$\Omega_T(t) = C_T G(\rho_{tot}(t)/R)\,\rho_{Mega} + \Omega_{Trest} \qquad (9.5)$$

where C_T is a constant, ρ_{tot} is the current total mass-energy density of the universe, and ρ_{Mega} is the total mass-energy density of the Megaverse and assumed to be constant.

9.3 Total Universe Mass-Energy Density

Taking account of the aforementioned features the total energy density of the universe at a given moment is

$$\rho_{tot}(t) \equiv \rho_{crit}\Omega_{tot}(t) = \rho_{crit}[\Omega_\Gamma(t) + \Omega_M(t) + \Omega_\Lambda + \Omega_{TST} + \Omega_{Trest}]$$

$$= \rho_{crit}[\Omega_\gamma(t) + \Omega_\Lambda + \Omega_m(t) + \Omega_d(t) + \Omega_{Trest} + C_T G\rho_{Mega}\int_{t_T}^{t} dt'\rho_{tot}(t')/R(t')\,]$$
$$(9.6)$$

---

[52] Callaway, D., Phys. Rev. E **53**, 3738 (1996).

where Ω_{Trest} is an additional possible energy density possibly due to quintessence, and assuming that the Megaverse mass-energy influx began at a time $t = t_T$.

In chapter 12 we will exhibit the time dependence of the quantities in eq. 9.6. and calculate the Hubble constant in chapter 19. Eq. 9.6 can be expressed as closed form solutions. Setting $\Omega_{Trest} = 0$ for the remainder of this section, we note

$$\rho_{tot}(t) = \rho_{crit}[\Omega_\gamma(t) + \Omega_\Lambda(t) + \Omega_m(t) + \Omega_d(t) + \Omega_{TST}(t)] \equiv \rho_{crit}[\Omega_H(t) + \Omega_{TST}(t)] \qquad (9.7)$$

We define

$$C_M = C_T G \rho_{Mega} \qquad (9.8)$$

and then take the derivative of eq. 9.6 to obtain

$$d\rho_{tot}(t)/dt = \rho_{crit}d\Omega_H(t)/dt + C_M \rho_{tot}(t)/R(t) \qquad (9.9)$$

with

$$\Omega_H(t) = \Omega_\gamma(t) + \Omega_\Lambda + \Omega_m(t) + \Omega_d(t) \qquad (9.10)$$

Solving eq. 9.9 we obtain

$$\rho_{tot}(t) = \rho_{crit}\exp\left[C_M\int_{t_T}^{t}dt'/R(t')\right] \int_{t_T}^{t}dt'' \exp\left[-C_M\int_{t_T}^{t''}dt'/R(t')\right] d\Omega_H(t'')/dt'' \qquad (9.11)$$

In chapter 14 we will calculate the total energy-momentum density as a function of time to obtain expressions for the scale parameter of the Robertson-Walker model and the resulting Hubble Constant.

10. Derivation of an Extended Robertson-Walker Model

This chapter provides the derivation of a generalization of the Roberson-Walker solution of General Relativity for use in our discussion of the Quantum Big Bang Theory originally presented in Blaha (2004). This chapter is extracted from chapter 8 of Blaha (2004).

The generalization derived here has not been derived before, to our knowledge, perhaps because, at the level of c-number General Relativity, it is fully equivalent to the known Robertson-Walker model. However when the theory has q-number parts introduced in a physically meaningful way, as we do, it is no longer equivalent to the c-number Robertson Walker model. The essential benefit of this quantized Robertson-Walker Model is that it supports a scale factor that is a function of both time and space. The quantum-based dependence on space (radial distance) enables the model to avoid infinities at $t = 0$.

10.1 The Robertson-Walker Metric

Much of the current modeling of the evolution and properties of the universe is based on the assumption of a Robertson-Walker metric which is used in the Einstein equations to obtain a first order differential equation for the scale factor $R(t)$:

$$\dot{R}^2 + k = 8\pi G\rho R^2/3 \qquad (10.1.1)$$

where k is a factor in the three-dimensional spatial curvature of the Robertson-Walker metric. The radius of curvature is

$$K_3(t) = k/R^2(t) \qquad (10.1.2a)$$

and the volume of the universe is

$$V = 2\pi^2 R^3$$

(10.1.2b)

Eq. 10.1.2 suggests that accurate measurements of the Hubble constant and other cosmological quantities could lead to an accurate determination of the curvature, the time dependence of the Hubble constant, and of R(t).

The form of the Robertson-Walker metric (We consider a real space-time only in this chapter) follows from the assumption of a maximally symmetric three-dimensional subspace whose metric has eigenvalues of the same sign (negative in our formalism) residing within a four-dimensional space-time with one positive eigenvalue and three negative eigenvalues. A maximally symmetric space is isotropic and homogeneous.[53]

Although the Robertson-Walker metric does not appear to embody the concept of an absolute space-time the general arguments presented in chapter 2 of Blaha (2004) indicate that it does in fact implicitly define an absolute reference frame.[54]

Therefore it is sensible to inquire whether a more general metric – a generalization of the Robertson-Walker metric – might be worth investigating – particularly in view of the major unexplained mysteries of Dark Energy as well as other new data showing the existence of massive black holes and quite mature galaxies shortly after (two or three billion years) the origin of the universe. The pile-up of mysteries from WMAP, SDSS and other sources indicates a reconsideration of fundamental assumptions may be worthwhile.

Therefore we will examine the "simplest" generalization of the Robertson-Walker metric in this chapter with a view towards elucidating some of these mysteries. More importantly, we will use this generalization in chapter 19 and subsequent chapters to develop a non-singular, Two-Tier formulation of the dynamics of the universe "at the beginning of time" – the Big Bang – taking account of quantum effects.

---

[53] See Chapter 13 of Weinberg (1972) for a detailed discussion.

[5454] This conclusion is now obvious due to the existence of the Megaverse. We note there can only be one absolute reference frame up to a Lorentz transformation since the sets of inertial reference frames of two absolute reference frames must be the same. This implies any two absolute reference frames must be related by a Lorentz transformation since absolute frames are necessarily flat inertial frames.

From the point of view of the definition of maximally symmetric subspaces the most immediate generalization of the Robertson-Walker metric is to assume a maximally symmetric *two-dimensional* subspace within a four-dimensional space-time. The general form of the metric in this case is:

$$d\tau^2 = A_{tt}(t, r) \, dt^2 + 2A_{rt}(r,t) \, dt \, dr + A_{rr}(t, r) \, dr^2 + B(r,t)(d\theta^2 + \sin^2\theta d\varphi^2) \quad (10.1.3)$$

where A_{ik} is a 2×2 symmetric matrix with one positive eigenvalue and one negative eigenvalue, and $B(r, t)$ is a negative function of r and t.

10.2 A Generalization of the Roberson-Walker Metric

We shall consider a generalization of the Robertson-Walker metric (eq. 5.5.1 of Blaha (2004)), which is a special case of eq. 10.1.3 that preserves the overall form of the Robertson-Walker metric but allows the scale factor a(t) to depend on r as well as t:

$$R(t) \to A_0(t, r) \qquad (10.2.1)$$

The generalized metric that we will analyze is embodied in the invariant interval expression:

$$d\tau^2 = dt^2 - A_0^2(t, r)[dr^2/(1 - kr^2) + r^2(d\theta^2 + \sin^2\theta \, d\varphi^2)] \qquad (10.2.2)$$

The introduction of a dependence on the radius r in $A_0(t, r)$ in eq. 10.2.2 eliminates the homogeneity of the three-dimensional spatial subspace reducing it to a two-dimensional maximally symmetric subspace.

We note that the general solution of the Einstein equations for the standard case of a perfect fluid lead to the usual view of the expansion of the universe (after the Big Bang Epoch), Hubble's law, the red shifts of radiation from distant sources of radiation, and the Cosmic Microwave Background (CMB) radiation.

10.3 The Einstein Equations for the Generalized Robertson-Walker Metric

The Einstein equations can be written

$$R_{\mu\nu} = -8\pi G S_{\mu\nu} \qquad (10.3.1a)$$

$$S_{\mu\nu} = T_{\mu\nu} - \tfrac{1}{2} g_{\mu\nu} T^{\sigma}{}_{\sigma} \qquad (10.3.1b)$$

where $R_{\mu\nu}$ is the Ricci tensor and $T_{\mu\nu}$ is the energy-momentum tensor. Assuming the energy-momentum tensor has the form of the energy-momentum tensor of a perfect fluid with the only non-zero components:

$$T_{tt} = \rho\, g_{tt} \qquad (10.3.2)$$

$$T_{rr} = -p g_{rr} \qquad (10.3.3)$$

$$T_{\theta\theta} = -p g_{\theta\theta} \qquad (10.3.4)$$

$$T_{\phi\phi} = -p g_{\phi\phi} \qquad (10.3.5)$$

where ρ is the density and p is the pressure, and the non-zero components of $S_{\mu\nu}$ are:

$$S_{tt} = \tfrac{1}{2}\,(\rho + 3p)\, g_{tt} \qquad (10.3.6)$$

$$S_{rr} = -\tfrac{1}{2}\,(\rho - p) g_{rr} \qquad (10.3.7)$$

$$S_{\theta\theta} = -\tfrac{1}{2}\,(\rho - p) g_{\theta\theta} \qquad (10.3.8)$$

$$S_{\phi\phi} = -\tfrac{1}{2}\,(\rho - p) g_{\phi\phi} \qquad (10.3.9)$$

In particular, the fact that

$$S_{tr} = 0 \qquad (10.3.10)$$

for a perfect fluid results in an important simplification in the solution of the Einstein equations for this case.

The density $\rho = \rho(t)$ and the pressure $p = p(t)$ are assumed to be solely functions of time t as is usual in the case of a perfect fluid.

10.4 Differential Equations for the Generalized Scale Factor A(t, r)

The dependence of the scale factor A_0 on both r and t leads to a significantly more complicated calculation of the Ricci tensor. We start by noting

$$g_{tt} = 1 \qquad g_{rr} = -A_0^2/(1 - kr^2) \qquad g_{\theta\theta} = -A_0^2 r^2 \qquad g_{\phi\phi} = -A_0^2 r^2 \sin^2 \theta \qquad (10.4.1)$$

Despite the dependence of A_0 on both t and r in the generalized case a direct calculation of the tt Ricci tensor component yields the familiar expression:

$$R_{tt} = g_{tt}\, 3\ddot{A}_0/A_0 \qquad (10.4.2)$$

where we use dots over A_0 to indicate partial derivatives with respect to time:

$$\ddot{A}_0 \equiv \partial^2 A_0/\partial t^2 \qquad (10.4.3)$$

However, the tr-component of the Ricci tensor R_{tr}, which is zero in the case of the ordinary Robertson-Walker metric, is non-zero in the more general case under consideration:

$$R_{tr} = 2\, \partial(\partial A_0/\partial t)/\partial r \qquad (10.4.4)$$

The corresponding Einstein equation is

$$R_{tr} = 2\, \partial(A_0^{-1}\, \partial A_0/\partial t)/\partial r = -8\pi G S_{tr} = 0 \qquad (10.4.5)$$

for a perfect fluid. Eq. 10.4.5 implies that $A_0(t, r)$ factorizes:

$$A_0(t, r) = a(t) b_0(r) \qquad (10.4.6)$$

This factorization results in a substantial simplification in the non-linear Einstein equations considered next, which are shown to be separable in the radial and time variables.

Before proceeding to the consideration of the remaining Einstein equations it is convenient to redefine the radial coordinate using

$$\check{r}\, b(\check{r}) = r\, b_0(r) \tag{10.4.7}$$

and

$$d\check{r}/dr = b_0(r)[b(\check{r})(1 - kr^2)^{\frac{1}{2}}]^{-1} = \check{r}\,[r(1 - kr^2)^{\frac{1}{2}}]^{-1} \tag{10.4.8}$$

While this change of coordinates does not change the physical content of the theory it does lead to simpler Einstein equations. The change of radial coordinate results in a new form of the invariant interval:

$$d\tau^2 = dt^2 - a^2(t)b^2(\check{r})[d\check{r}^2 + \check{r}^2(d\theta^2 + \sin^2\theta\, d\varphi^2)] \tag{10.4.9}$$

The new radial coordinate $\check{r}$ is related to the old radial coordinate by

$$\check{r} = \{[1 - (1 - kr^2)^{\frac{1}{2}}]/[1 + (1 - kr^2)^{\frac{1}{2}}]\}^{\frac{1}{2}} \tag{10.4.10}$$

and

$$r = 2k^{-\frac{1}{2}}\check{r}(1 + \check{r}^2)^{-1} \tag{10.4.11}$$

Note the range of $\check{r}$ is [0, 1].

A direct calculation of the Ricci tensor for the metric in eq. 10.4.9 with $A(t, \check{r})$ defined as:

$$A \equiv A(t, \check{r}) = a(t)b(\check{r}) \tag{10.4.12}$$

leads to the following Einstein equations (remembering our flat space Cartesian metric is $\eta_{tt} = +1$ and $\eta_{ij} = -\delta_{ij}$ for i, j = spatial indices):

$$R_{tt} = g_{tt}3\ddot{A}/A = -8\pi GS_{tt} = -4\pi G(\rho + 3p) \tag{10.4.13}$$

$$R_{t\check{r}} = 2\, \partial(A^{-1}\, \partial A/\partial t)/\partial\check{r} = -8\pi G\, S_{t\check{r}} = 0 \tag{10.4.14}$$

$$R_{\check{r}\check{r}} = g_{\check{r}\check{r}}[\ddot{A}/A + 2(\dot{A}/A)^2 - 2(\check{r}A^2)^{-1}\, \partial(\check{r}\, A'/A)/\partial\check{r}]$$

$$= -8\pi G \; S_{\check{r}\check{r}} = 4\pi G \; g_{\check{r}\check{r}}(\rho - p) \tag{10.4.15}$$

$$R_{\theta\theta} = g_{\theta\theta}[\ddot{A}/A + 2(\dot{A}/A)^2 - A''/A^3 - 3A'/(\check{r}A^3)^{-1}]$$

$$= -8\pi G \; S_{\theta\theta} = 4\pi G \; g_{\theta\theta}(\rho - p) \tag{10.4.16}$$

$$R_{\phi\phi} = g_{\phi\phi}[\ddot{A}/A + 2(\dot{A}/A)^2 - A''/A^3 - 3A'/(\check{r}A^3)^{-1}]$$

$$= -8\pi G \; S_{\phi\phi} = 4\pi G \; g_{\phi\phi}(\rho - p) \tag{10.4.17}$$

where

$$A' = \partial A / \partial \check{r} \tag{10.4.18}$$

and

$$A'' = \partial^2 A / \partial \check{r}^2 \tag{10.4.19}$$

10.5 The Solution for the Generalized Scale Factor A(t, r)

We begin by substituting eq. 10.4.13 in eq. 10.4.15, and then substituting the factorization of A (eq. 10.4.12). The result is a separable equation:

$$\dot{a}^2 - 8\pi G\rho a^2/3 - (\check{r}b^2)^{-1} \; \partial(\check{r} \; b'/b)/\partial\check{r} = 0 \tag{10.5.1}$$

Since the first two terms in eq. 10.5.1 are solely functions of t while the third term is solely a function of $\check{r}$ we obtain the separated equations:

$$\dot{a}^2(t) - 8\pi G\rho a^2(t)/3 = -k \tag{10.5.2}$$

where $\rho = \rho_{tot}$ and

$$k + (\check{r}b^2)^{-1} \; \partial(\check{r} \; b'/b)/\partial\check{r} = 0 \tag{10.5.3}$$

where k is a separation constant, which we provisionally identify as the curvature parameter k in eq. 10.2.2.

Eq. 10.5.2 is precisely the equation used for the time dependent scale factor in current cosmological models (eq. 5A.3.2) using the Robertson-Walker metric. Therefore its solution, which depends on the time dependence of the energy density, will be the same as that of the corresponding conventional cosmological model for the same density.

Eq. 10.5.3 is a differential equation for the *spatial* expansion scale factor $b(\check{r})$. Under the assumption of a perfect fluid it depends solely on the constant k and is independent of the details of the perfect fluid (i.e. its density and pressure). There are two solutions of the second order non-linear differential equation for $b(\check{r})$. These solutions can be written as

$$b_1(\check{r}) = 2\gamma\delta k^{-\frac{1}{2}}[\delta^2 + \gamma^2\,\check{r}^2]^{-1} \tag{10.5.4}$$

where γ and δ are constants, and

$$b_2(\check{r}) = \sigma k^{-\frac{1}{2}}[\check{r}\,(\varsigma \pm i\sigma \ln \check{r})]^{-1} \tag{10.5.5}$$

where σ and ς are constants.

$b_1(\check{r})$ (in eq. 10.5.4) is the only physically acceptable solution for the case of a perfect fluid with energy-momentum tensor specified by eqns. 10.3.2 – 10.3.5. Reason: The solution $b_1(\check{r})$ satisfies eqns. 10.4.16 and 10.4.17 while the other solution $b_2(\check{r})$ does not satisfy these equations. A necessary and sufficient condition for $b(\check{r})$ to satisfy eqns. 10.4.16 and 10.4.17 is that

$$b'' - \check{r}^{-1}\,b' - 2\,b'^2/b = 0 \tag{10.5.6}$$

where $'$ denotes a derivative with respect to $\check{r}$. Eq. 10.5.6 follows from subtracting the coefficients of the metric tensor component factors in eq. 10.4.15 from the coefficients of the metric tensor component factors in eq. 10.4.16 (or eq. 10.4.17). The solution $b_1(\check{r})$ satisfies eq. 10.5.6 for all values of γ and δ. The solution $b_2(\check{r})$ does not satisfy eq. 10.5.6 for any choice of α and β except the trivial choice $\alpha = 0$ and is therefore physically irrelevant. It might have some relevance in the case of a non-perfect fluid. We will not investigate that possibility in the present work.

10.6 The Solution Expressed in the Original Radial Coordinate

The solution that we have obtained for the generalized Roberson-Walker case with radial coordinate ř can be related back to the generalized Robertson-Walker solution using the original radial coordinate r. Eq. 10.4.7 implies

$$b_0(r) = ř\, b(ř)/r \qquad (10.6.1)$$
$$= 2a\,[1 + a^2 + (1 - a^2)(1 - kr^2)^{\frac{1}{2}}]^{-1} \qquad (10.6.2)$$

using eqns. 10.4.10 and 10.5.4, and defining the constant α as

$$\alpha = \gamma/\delta \qquad (10.6.3)$$

An important special case of eq. 10.6.2 is the case where $a = 1$ (Ockham's Razor!). In this case we find eq. 10.6.2 becomes

$$b_0(r) = 1 \qquad (10.6.4a)$$

and

$$A_0(t, r) = a(t) \qquad (10.6.4b)$$

thus recovering the normal Robertson-Walker solution exactly as a special case from eqns. 10.2.2, 10.4.6, and 10.5.2. In this case we see

$$b(ř) = 2k^{-\frac{1}{2}}(1 + ř^2)^{-1} \qquad (10.6.5)$$

from eq. 10.5.4 in the ř, θ, ϕ coordinate system. We considered this case within the expanded framework of a quantized model of the beginning of the universe in chapter 19 where it has non-trivial consequences.

10.7 Equivalence of the General Solution with the Original Robertson-Walker Solution

The general solution of the Einstein equations in the case of a perfect fluid (eqns. 10.4.13 – 10.4.17) for the scale factor $A(t, \check{r})$ in the generalized metric specified by eq. 10.4.9

$$d\tau^2 = dt^2 - A^2(t, \check{r})[d\check{r}^2 + \check{r}^2(d\theta^2 + \sin^2\theta \ d\varphi^2)] \tag{10.4.9}$$

is given by

$$A(t, \check{r}) = a(t)b(\check{r}) = 2ak^{-\frac{1}{2}}a(t)[1 + a^2\check{r}^2]^{-1} \tag{10.7.1}$$

where a is a constant given by eq. 10.6.3 and where $a(t)$ is the solution of the standard equation (eq. 10.5.2) for the time dependent scale factor in the case of the Robertson-Walker metric. We now note that if we define a new radial vector

$$\mathfrak{r} = a\check{r} \tag{10.7.2}$$

then

$$d\tau^2 = dt^2 - a^2(t)4k^{-1}[1 + \mathfrak{r}^2]^{-1} \ [d\mathfrak{r}^2 + \mathfrak{r}^2(d\theta^2 + \sin^2\theta \ d\varphi^2)] \tag{10.7.3}$$

Comparing this invariant interval expression with the $a = 1$ expression for $b(\check{r})$ given in eq. 10.6.5 we conclude that we have proved the following theorem:

Theorem: The solution of the Einstein equations for the case of a perfect fluid for the generalized Robertson-Walker metric (eq. 10.2.2) is equivalent to a solution of the Einstein equations for the case of a perfect fluid in the case of the Robertson-Walker metric with a scale factor $a(t)$ that is solely dependent on time.

In the case of *classical* (non-quantum) gravitation theory the solutions are fully equivalent and related by a simple change of radial coordinates. Thus the homogeneity condition that we had relinquished at the beginning of our discussion is reinstated and the solution of the generalized case is consistent with a *maximally symmetric three-*

dimensional space. The origin of the coordinate system can be chosen to be any point in space.

In the case of quantized versions of the Robertson-Walker model and our generalization of it we will see that the solutions are *generally not equivalent*. We explored a particular example of a quantized gravitational model that illustrates this point in chapter 8. The quantized model, which we defined, should be viewed as a first attempt to explore the quantum regime existing at the beginning of time – the Big Bang. The reasonableness of its results suggests that we are on the right track for understanding the Big Bang Epoch.

10.8 Form of Hubble's Law in the Generalized Robertson-Walker Model

Our generalized Robertson-Walker metric assumes an inhomogeneous space with some fixed center at $\check{r} = 0$. Presumably this center was the point at which the Big Bang took place at the beginning of the universe. In this section see how Hubble's Law emerges in the generalized model.

Hubble's law is one of the cornerstones of modern cosmology. While one might think a scale factor that depended on the radius coordinate might not be consistent with Hubble's Law it is easy to show that Hubble's Law is satisfied provided that the scale factor factorizes as required by the tr-component of Einstein's equations (eq. 10.4.14) for our generalized Robertson-Walker model.

First we give a simple derivation of Hubble's law for the case of a separable scale factor:

$$A(t, \check{r}) = a(t)b(\check{r}) \qquad (10.4.12)$$

under the assumption that some remote galaxy lies on the same radial line as the line from the origin of the space coordinates to our galaxy. The proper distance between the remote galaxy and our galaxy has the form

$$D(t) = D_0 A(t, \check{r}) \qquad (10.8.1)$$

The rate of recession of the remote galaxy is then

$$v = dD/dt = D_0 b(\v{r})da(t)/dt = HD(t) \tag{10.8.2}$$

with

$$H = d(\ln a(t))/dt \tag{10.8.3}$$

H is the Hubble Constant. If the speed of recession is small then it determines the first-order Doppler shift. If we denote the wavelength of the received radiation as λ_r and the wavelength of the radiation at the source as λ_s then the shift z is

$$z = \lambda_r/\lambda_s - 1 = v/c = HD/c \tag{10.8.4}$$

Eq. 10.8.4 is Hubble's Law: the red shift of a galaxy is proportional to its distance. Note Hubble's Law in the generalized case follows from the factorization of $A(t, \v{r})$ with $a(t)$ satisfying the same differential equation as the Robertson-Walker scale factor.

Since a change of radial coordinate reduces the classical generalized model to the Robertson-Walker model, Hubble's Law can be proven in the general case of the non-collinearity of the coordinate origin, source and reception points.

11. Extended Robertson-Walker Model with Quantum Coordinates

In this chapter we introduce a form of quantum coordinates (called Two-Tier coordinates) into the Extended Robertson-Walker Model that we created in chapter 10. Its role is to eliminate the infinities at the Big Bang point. It has the remarkable feature that its effect becomes negligible shortly (within nanoseconds) after the Big Bang. Thus the form of the expansion scale factor is virtually identical to that of the usual Robertson-Walker scale factor. In Blaha (2018e) we showed there are a number of other important reasons for introducing our Quantum Coordinates beyond the problems with the Big Bang:[55]

"Originally Two-Tier coordinates were developed by this author to remove infinities that appear in perturbation theory calculations. We showed that the quantum smeared coordinates of Two-Tier Quantum Field Theory succeeded in removing all ultra-violet infinities in perturbation theory including the fermion triangle infinities. Remarkably the high precision, low energy[56] predictions of QED remained true in Two-Tier QED and thus remained consistent with experiment to a hitherto unsurpassed level of accuracy. 'Low' energy predictions in other quantum field theories also remained unchanged. At high energies, Two-Tier perturbation theory results are finite and consequently all ultra-violet infinities, to any order in perturbation theory, in *any number of space-time dimensions* were eliminated.

In addition to removing perturbation theory infinities Two-Tier coordinates enable us to define finite theories of Quantum Gravity and 'non-renormalizable' quantum field theories based on polynomial lagrangians, to tame vacuum fluctuations, to eliminate infinities associated with the Big Bang, and possibly to generate the explosive growth of the universe in its role as Dark Energy.[57]

Two-Tier Quantum Field Theory is established on the most fundamental level.
    …

---

[55] Extract from Blaha (2018e).
[56] Relative to a mass scale that was perhaps of the order of the Planck mass.
[57] See Blaha (2017b) and earlier books for details. This section is basically a summary of some features.

Thus using Two-Tier Quantum Field Theory we can perform perturbation theory calculations that always yield a finite result in any number of dimensions.[58] This is not true if conventional Quantum Field is used.[59]"

In Blaha (2018e) we derived the form of the Unified SuperStandard Theory. It includes Gravitation, Dark Matter, a fourth fermion generation, and four layers of four generation fermions totaling 192 fermions. Similarly we found 192 non-abelian vector interaction bosons as well as a U(4) General Relativity-based non-abelian vector boson and a U(192) interaction rotations group.

We introduced a universal field Y^μ that eliminated the divergences that appear in perturbation theory without the use of a renormalization program. We now turn to the extension of our theory to the Cosmology of our universe, and the extension of Nature to include other universes, which together with our universe, reside in a D dimensional Megaverse.

It seemed appropriate to study the evolution of our universe from its beginning, and view it as a model of the general evolution of universes in the Megaverse. There does not appear to be any reason to view our universe as unique. Given the oft-noted tendency of Nature to repeat the characteristics of phenomenon we believe our universe should provisionally be viewed as a 'typical' universe.

Almost all of this chapter,[60] and the following chapters, first appeared in Blaha (2004) to derive our Quantum Big Bang Theory. The new aspect is its unification with a Steady State Theory of the Hoyle-Narlikar type to create a hybrid Big_Bang-Steady_State Theory.[61] Unlike Hoyle and Narlikar we can attribute the appearance of mass-energy in the universe to an influx from the Megaverse using our surface tension force concept.

---

[58] In particular, the fermion triangle divergence (anomaly) does not occur in our Two Tier Quantum Field Theory of the fermion sector. Thus there is no requirement for axion-like particles in the Megaverse (or in universes) although the possible existence of this type of particle is not ruled out.

[59] Blaha (2005a) provides a complete discussion of Two-Tier Quantum Field Theory.

[60] While the calculation is unchanged we have added clarifications of the original that might have led to misunderstandings particularly in clarifying the *constant* α vs. the scale function a(t).

[61] In section 3 we abandon the Steady State part of this theory in favor of a theory combining our Big Bang theory with a universal scale parameter theory which we justify in section 4 as a consequence of universe vacuum polarization.

11.1 Dark Energy

A form of Dark Energy surfaced in Cosmology when it was noticed that the speed of expansion of our universe was expanding due to an unknown, and "undetectable" energy, dubbed Dark Energy. A theoretical framework was developed by A. Guth and others based on an *inflation* that provided a scenario for the expansion of the universe through an unidentified source called Dark Energy.

We shall now show that one cause of inflatons is the $Y^\mu(y)$ quantum field of our Extended Standard Model that also resolved QFT divergence problems. We will see $Y^\mu(y)$ also makes the Big Bang finite—no singularities; as well as freeing The Standard Model and Quantum Gravity[62] of infinities. *Thus $Y^\mu(y)$ has a remarkable triple role in our view – to eliminate the Big Bang singularity, to generate the explosive growth of the universe, and to remove infinities from The Standard Model and Quantum Gravity.* This happy coincidence of solutions reflects an Ockham's Razor approach to find the simplest general solution and Leibniz's Minimax Principle to find the most minimal solution that has maximal effects – two sides of the same coin in a sense.

Since the $Y^\mu(x)$ quantum gauge field is a free field (neglecting gravity) the initial state of the universe can be permeated with quanta of this field as well as particle quanta. We call the total energy of the free $Y^\mu(x)$ field within the universe Y Dark Energy.

11.2 The Big Bang Experimentally

In the light of our progress since 2004, we now provide a revised version that is quite similar to Blaha (2004).

The current state of our knowledge of the evolution of the universe has now been extended back in time to about 380,000 years after the Big Bang through recent astrophysical research. While this progress is encouraging we still face major issues: the nature of Dark Matter (hopefully resolved by the Unified SuperStandard Theory), the nature and origin of Dark Energy (hopefully resolved in this chapter), and the events of those critical years before the 350,000 year point that we are slowly reaching

---

[62] See Blaha (2011c) and Blaha (2005a) for the removal of infinities in Quantum Gravity.

experimentally. Those early years and the Big Bang itself were mysteries. This situation was especially critical since the early years of the universe apparently contain an uncertain beginning and an explosive growth.

In this chapter we will attempt to understand that unknown period in the neighborhood of t = 0 where we believe quantum coordinate effects play a major role. We will suggest that the inflationary growth of the universe, which is attributed to an unknown "particle", actually is caused by the energy of the q-number part of coordinates – the quantum field $Y^\mu(x)$ that we saw in earlier chapters.

11.3 The State of the Universe at t = 0

If we simply extrapolate the currently popular Standard Cosmological Models (with or without inflations) back to the Big Bang t = 0, we find a universe beginning as a single "mathematical point" with infinite mass density and infinite temperature. The Robertson-Walker metric scale factor $a(t) \equiv R(t)k^{\frac{1}{2}}$, which is a solution of the Einstein equation

$$\dot{a}^2 - 8\pi G\rho_{tot}a^2/3 = -k \qquad (11.1)$$

typically is solved for a perfect fluid under the assumption of a matter-dominated, or a radiation-dominated, phase of the universe. If we assume the universe is matter-dominated, then the energy density is

Matter-Dominated: $\qquad \rho = \rho_0/a(t)^3 \qquad (11.2)$

Under the alternate assumption that the universe is radiation-dominated we have[63]

Radiation-Dominated: $\qquad \rho = \rho_0/a(t)^4 \qquad (11.3)$

With either assumption we find that the scale factor behaves as

---

[63] *When we consider the mass-energy influx from the Megaverse we will assume the influx is primarily radiation just like in our universe.*

$$a(t) \propto t^n \tag{11.4}$$

where $0 < n < 1$. Thus $a(0) = 0$, and the universe reduces to a point with infinite density (eqns. 11.2 and 11.3), and with infinite temperature since

$$T \propto a^{-1}(t) \tag{11.5}$$

There are evidently grave difficulties in extrapolating the Standard Cosmological Model, or its current variants, to $t = 0$. The difficulty is compounded by the inherently quantum mechanical aspects that are normally associated with gravitation at ultra-small distances.

11.3.1 Two-Tier Quantum Coordinates in a Quantum Gravity

Currently, the only viable complete theory of Quantum Gravity is the Two-Tier Quantum Gravity of Blaha (2004) and (2005a). Blaha's type of quantum field theory has the interesting feature that all forces (particle propagators) become zero at very short distances (presumably much less than the Planck mass). Thus a point universe could have an infinite density of essentially "non-interacting" matter as a quasi-stable state. Furthermore if one uses the Extended Robertson-Walker metric as described in Blaha (2004), and Chapter 10 of this book, one finds a classical scale factor of the form:

$$A(t, \v{r}) = a(t)b(\v{r}) \tag{11.6}$$

where $a(t)$ satisfies eq. 11.1.

We now assume $A(t, \v{r})$ is a quantum operator due to the Two-Tier quantization of $\v{r}$ as shown in chapter 7. *The relation of the R scale factor, and other forms of the scale factor, is summarized by*

$$R(t) = k^{-\frac{1}{2}}a(t) \rightarrow A_0(t, r) \tag{10.2.1}$$
$$A_0(t, r) = a(t)b_0(r) \tag{10.4.6}$$
$$A \equiv A(t, \v{r}) = a(t)b(\v{r}) \tag{10.4.12}$$

Since quantum effects will play a role near $t = 0$ (the "Big Bang") it is likely that the expectation value of $A(t, \v{r})$ will have the form

$$<A(t, \check{r})> = <a(t)><b(\check{r}, t)> \tag{11.7}$$

where

$$<b(\check{r}, t)> \to \beta(\check{r})/<a(t)> \tag{11.8}$$

as $t \to 0$. Then the zero of $<a(t)>$ might be cancelled with the result

$$<A(t, \check{r})> \to \beta(\check{r}) \neq 0 \tag{11.9}$$

Quantum effects would thus eliminate the singularities at $t = 0$. A quantized version of the generalized Robertson-Walker model[64] opens the possibility of a universe with a finite size, density, and temperature at the time of the Big Bang.

With that possibility in mind, we will use an extension of Blaha (2004)'s Two-Tier theory to develop an enhanced version of his quantum model of the universe in the neighborhood of $t = 0$. Starting from eq. 11.7.1 of Blaha (2004) which is eq. 10.7.1 in Chapter 10:

$$A(t, \check{r}) = a(t)b(\check{r}) = 2\alpha k^{-\frac{1}{2}}a(t)[1 + \alpha^2 \check{r}^2]^{-1} \tag{11.7.1}$$

where α is a constant[65] specified by eq. 10.6.3 that we introduce a Two-Tier variable Y^μ (as in chapter 7 with the identification[66]

$$\check{r} \equiv M_c X = M_c(y + iY/M_c^2)$$

where $Y = |\mathbf{Y}|$ we see

$$b(\check{r}) = b(y, t) = 2\alpha k^{-\frac{1}{2}}[1 + \alpha^2(M_c y + iY/M_c)^2]^{-1} \tag{11.10}$$

If (as seen later)

$$Y = -M_c[a_1(y, t) - a_2(y, t)a(t)]^{\frac{1}{2}}/\alpha + iM_c^2 a_3(y, t) \tag{11.11}$$

---

[64] This model appears in Blaha (2004).

[65] The constsant α is not time dependent; a(t) is time dependent. The lack of an argument further distinguishes α from a(t).

[66] α is a constant and not the fine structure constant. We set a = 1 later.

and if, as $t \to 0$,

$$a_1(y, t) \to a_1(y, 0) = 1 \qquad (11.12)$$
$$a_2(y, t) \to a_2(y, 0) \neq 0 \qquad (11.13)$$
$$a_3(y, t) \to a_3(y, 0) = y \qquad (11.14)$$

then we find

$$b(y, t) \to 2ak^{-1/2}/[a_2(y, 0)a(t)] \qquad (11.15)$$

and

$$<A(t, \check{r})> = <a(t)b(\check{r})? = 2ak^{-1/2}/a_2(y, 0) + a(t)\beta_1(y) + a^2(t)\beta_2(y) + \ldots \qquad (11.16)$$

$$\to 2ak^{-1/2}/a_2(y, 0) \qquad \text{as } t \to 0$$

Thus, under these circumstances, space does not collapse to a point, and the density and temperature—as well as other parameters of interest—are finite. *Yet the features of the Standard Cosmological Model at larger times still remain valid.*

11.3.2 Assumptions of Our Two-Tier Quantum Theory

In our model we will make the following assumptions about the universe near $t = 0$:

1. The particles in the universe consist of fundamental elementary particles – gravitons, photons, electrons, neutrinos, quarks, gluons and so on – and their corresponding anti-particles.

2. The particles are described by Two-Tier quantum field theory. In this type of quantum field theory all particle interactions become negligible at very short distances (as described in chapter 7 of Blaha (2004)) and so the forces between particles may be neglected near $t = 0$ when the universe is immensely *small*.

3. The energy of the universe can be viewed as consisting of particles – bosons and fermions – each species having blackbody energy distributions since the universe is the best of all possible black bodies.

4. The enormous energy of the universe, even if confined to a small region, makes the classical Einstein equations a good approximation *due to its macroscopic nature* with one proviso (item 5). Therefore we assume the Extended Robertson-Walker metric of Blaha (2004) and chapter 10.

5. In the neighborhood of t = 0 when the universe is effectively confined to a region whose scale is set by the Planck mass or smaller the quantum nature of the Two-Tier coordinate X^μ becomes significant. In particular the Y^μ field causes a profound change in the behavior of the scale factor A(t, r) as t → 0. (Note: $X^\mu(y) = y^\mu + iY^\mu(y)/M^2$ defines the quantum coordinates where y^μ is a c-number coordinate and Y^μ a free q-number field similar to the electromagnetic field.)

6. The Y^μ quanta are assumed to have a black body spectrum[67] – just like elementary particles – reflecting their continuous emission and absorption by gravitons and other elementary particles. The Y^μ blackbody spectrum is implemented via a coherent state. Effectively the coherent state opens a small "bubble" into complex space-time changing the dynamics of the universe at t = 0.

7. We will calculate the expectation value of the quantum field operator Y^μ in a closed Robertson-Walker space. In principle we must use the Extended

---

[67] This is the only reasonable choice for the spectrum is a black body spectrum given the confinement of the field to the ultimate black body – the universe.

Robertson-Walker metric since the scale factor will depend on both r and t through its dependence on the expectation value of Y^μ.

11.4 Two-Tier Quantum Hybrid Model for the Beginning of the Universe

Our approach in this, and the following, sections will follow a modest program using the known theoretical foundations of elementary particle physics: the Extended Standard Model unified with Quantum Gravity in a Two-Tier quantum field theoretic framework. We will supplement this framework with natural assumptions about the initial conditions of the universe in order to develop a theory describing the evolution of the universe from its initial state.

11.4.1 Einstein Equations Near t = 0

There is no physical reason to believe that the universe at the beginning of time, t = 0, was a mathematical point of infinite temperature and density since the extrapolation of the scale factor of the Standard Cosmological Model to t = 0 is unwarranted for many reasons including quantum considerations.

Ideally we would use the Quantum Theory of Gravity to establish the physical theory of the universe near t = 0. However a quantum calculation of the global structure of the universe near t = 0 is not feasible. In view of this situation we must find an approximation that captures the physics of the universe near the Big Bang. One approach is based on the macroscopic energy of the early universe. One can expect that a classical gravitation model with appropriate quantum corrections may be a reasonable approximation to the early state of the universe. After all, macroscopic bodies are described by classical physics in general. And the universe is a macroscopic body by virtue of its content at the point of the Big Bang despite its small size. Therefore we will assume that we may start with a classical gravitation model and then introduce quantum effects.

The natural first choices – based on symmetry considerations – are a Robertson-Walker model and the Extended Robertson-Walker model described in chapter 8 of Blaha (2004) and Chapter 10 here. The quantum part, that we will shortly introduce, will require us to use an Extended Robertson-Walker model since quantum corrections reduce the symmetry to a maximally symmetric *two-dimensional* subspace within a four-dimensional space-time. *The quantum part eliminates the equivalence of the classical Robertson-Walker model and the Extended Robertson-Walker models* that was described in section 8.7 of Blaha (2004).

       Therefore we begin with the classical c-number equation for the invariant interval defined by

$$d\tau^2 = dt^2 - A^2(t, \check{r})[d\check{r}^2 + \check{r}^2(d\theta^2 + \sin^2\theta \, d\varphi^2)] \tag{11.17}$$

where

$$A(t, \check{r}) = a(t)b(\check{r}) = 2\alpha k^{-\frac{1}{2}}a(t)[1 + \alpha^2 \, \check{r}^2]^{-1} \tag{11.18}$$

is a q-number and where α is a constant[68] given by eq. 10.6.3. The scale factor $a(t)$ is the solution of the Einstein equation:

$$\dot{a}^2(t) - 8\pi G\rho_{tot}a^2(t)/3 = -k \tag{11.19}$$

Next we introduce quantum coordinates

$$X^\mu = y^\mu + i \, Y^\mu(y)/M_c^2 \tag{11.20}$$

We choose the same transverse gauge for Y^μ as we did in chapter 7 of Blaha (2004):

$$\partial Y^i/\partial y^i = 0 \tag{11.21}$$

$$Y^0 = 0 \tag{11.22}$$

As a result we make the identification (definition of coordinates)

$$X^0 = y^0 \equiv t \tag{11.23}$$

$$X^j = y^j + i \, Y^j(y)/M_c^2 \equiv M_c^{-1}\check{r}^j \tag{11.24}$$

---

[68] We use α to denote a constant, later set equal to 1, and $a(t)$ to denote the solution of the Einstein equation above. They are not connected to each other.

The mass factor on the right side of the equal sign in eq. 11.20 is required on dimensional grounds if y is to have the usual dimension of length (inverse mass). As a result, since $\check{r} \in [0, 1]$ by Blaha (2004), eq. 11.24 implies

$$y = |\mathbf{y}| \in [0, M_c^{-1}] \qquad (11.25)$$

There are two constants with the dimension of mass to a power: k and M_c. The constant k determines the curvature of space – a large-scale feature of Robertson-Walker models. The constant M_c is related to the very short distance behavior of the theory – high energy phenomena with energies of the order of the Planck mass or larger, and, as we will see, the origin of the universe – a short distance, high energy phenomena as well. Therefore we have also chosen to use M_c on the right side of eq. 11.24.

Since X^0 is a c-number and since the density $\rho(t)$ is a large c-number to very good approximation we will assume a(t) is the c-number solution of the classical c-number eq. 11.19 as $t \to 0$. Further we assume that quantum effects appear solely through $b(\check{r})$. We also assume that the q-number equivalent of the c-number $b(\check{r})$ is $b(M_cX)$. The q-number function $b(M_cX)$ satisfies the functional equation:[69]

$$k + (M_c^2 X b^2)^{-1} \, \partial(X b'/b)/\partial X = 0 \qquad (11.26)$$

where

$$b' = \partial b/\partial X \quad b = b(M_cX) \qquad (11.27)$$

and $X = (\mathbf{X \cdot X})^{\frac{1}{2}}$. The formal solution of eq. 11.26 has the same functional form as the c-number solution $b(\check{r})$ in eq. 10.5.4. Therefore

$$b(M_cX) = :2\alpha k^{-\frac{1}{2}}[1 + \alpha^2 M_c^2 X^2]^{-1}: \qquad (11.28)$$

where α is the constant specified by eq. 10.6.3 and where we have specified normal ordering with colons : ... : to avoid trivial divergences.

---

[69] See eq. 10.5.3 for the c-number equivalent.

Since eq. 11.28 is a q-number expression we must find the scale factor as the expectation value of A(t, X) for a suitable state. We note that, at this point, the invariant interval is an operator expression of the form:

$$d\tau^2 = dt^2 - B^2(t, X)[dX^2 + X^2(d\theta^2 + \sin^2\theta \, d\varphi^2)] \qquad (11.29)$$

where

$$B(t, X) = M_c A(t, M_c X) = a(t)b_M(X) \qquad (11.30)$$

and

$$b_M(X) = M_c b(M_c X) = \, :2aM_c k^{-\frac{1}{2}}[1 + a^2 M_c^2 X^2]^{-1}: \qquad (11.31)$$

where α is a constant (eq. 10.6.3). Thus the expectation value of $b_M(X)$ also must be calculated in order to determine the invariant interval's expectation value.

11.4.2 Y Black-Body Coherent States

The Y quanta are continuously being emitted and absorbed by the particles in the primeval universe. As such, they may be expected to have a blackbody energy spectrum that is similar to that of the particles from which they derive their existence. In particular one expects the temperature T associated with their blackbody energy distribution to be the same as that of the "real" particles in the universe. After all, the universe is a type of black body.

Thus the blackbody energy of Y-quanta as a function of frequency v per unit volume per unit frequency is assumed to be:

$$u_v = 8\pi h c^{-2} v^3 \, [e^{hv/\kappa T} - 1]^{-1} \qquad (11.32)$$

where c is the speed of light, h is Planck's constant, and κ is Boltzmann's constant. At this point we adopt units in which c = 1 and $\hbar = 1$.

The Hamiltonian for the Y field has a form that is familiar from electrodynamics

$$H = \int d^3y \, \mathcal{H}_Y(y) = \tfrac{1}{2}\int d^3y \, :E_Y^2 + B_Y^2: \, = \int d^3p \, \omega \sum_\lambda a^\dagger(p,\lambda)a(p,\lambda) \qquad (11.33)$$

where E_Y and B_Y are the "electric" and "magnetic" parts of the Y field, $\omega = p^0 = |\vec{p}| = 2\pi v$ is the energy (in our units), and λ labels the polarization. Note that we are using "infinite volume" continuum quantization formulation.

We now define coherent Y field bra and ket states that yield a spherically symmetric blackbody distribution as the eigenvalue of the Hamiltonian H:

$$|BB, T> = N \exp[\int d^3p \ f(\omega,T) \sum_\lambda a^\dagger(p, \lambda)]|0> \qquad (11.34)$$

$$<BB, T| = N^* <0|\exp[\int d^3p \ f^*(\omega,T) \sum_\lambda a(p, \lambda)] \qquad (11.35)$$

where $\omega = |\vec{p}|$, and where N is a normalization factor. The expectation value of H is

$$<BB, T|H|BB, T> = \int d^3p \ 2\omega|f(\omega,T)|^2 \qquad (11.36)$$

$$= \int d\omega \ 8\pi\omega^3|f(\omega,T)|^2 \qquad (11.37)$$

$$= \int dv \ 16\pi^2\omega^3|f(\omega,T)|^2 \qquad (11.38)$$

where $\omega = 2\pi v$ in our units ($c = 1$, $\hbar = 1$), and where the factor of two in eq. 11.36 is the number of polarizations.

The expectation value (eigenvalue) of the energy per unit frequency is

$$H_v = 16\pi^2\omega^3|f(\omega,T)|^2 \qquad (11.39)$$

We relate H_v to the blackbody energy *per unit volume* per unit frequency u_v using

$$H_v = u_v(2\pi/\omega)^3 \qquad (11.40)$$

where the factor of $(2\pi/\omega)^3$ makes the right side of eq. 11.40 the blackbody energy per unit frequency in the continuum case of a quantum field in a space of infinite volume. Thus we find

$$f(\omega,T) = \omega^{-3/2}[e^{\omega/\kappa T} - 1]^{-1/2} \qquad (11.41)$$

with the phase of $f(\omega,T)$ set to zero.

11.4.3 Expectation Value of Y in Coherent States

As a preliminary to the evaluation of the operator scale factor in eq. 11.31 we will evaluate the expectation value of powers of the Y field between black body coherent states defined by eqns. 11.34-11.35. We will then determine the expectation value of $b_M(X)$ to obtain the behavior of the overall scale factor near t = 0. It should be apparent to the reader that the expectation value of $b_M(X)$ is dependent on t as well as y due to the time dependence of the Y field. Thus the scale factor will exhibit a considerably more intricate behavior than simply its a(t) dependence.

The Fourier expansion of the Y field is:

$$Y^i(z) = \int d^3p \, N_0(\omega) \sum_{\lambda=1}^{2} \varepsilon^i(p, \lambda)[a(p,\lambda) \, e^{-ip\cdot z} + a^\dagger(p,\lambda) \, e^{ip\cdot z}] \qquad (11.42)$$

for i = 1, 2, 3 where z^μ will be set equal to y^μ later, and where

$$N_0(\omega) = [(2\pi)^3 2\omega]^{-\frac{1}{2}} \qquad (11.43)$$

and

$$\omega = (\mathbf{p}^2)^{\frac{1}{2}} = p^0 \qquad (11.44)$$

with $\vec{\varepsilon}(p, \lambda)$ being the polarization unit vectors for $\lambda = 1, 2$ and $\eta_{\mu\nu}p^\mu p^\nu = 0$. The expectation value of Y between the |BB, T> states is:

$$<BB, T|Y^i(z)|BB, T> = \int d^3p \, N_0(\omega)f(\omega,T)[e^{-ip\cdot z} + e^{ip\cdot z}] \sum \varepsilon^i(p, \lambda) \qquad (11.45)$$

The evaluation of eq. 11.45 (and spherical symmetry) gives

$$<BB, T|Y^i(z)|BB, T> = \hat{y}^i \int d^3p \, N_0(\omega)f(\omega,T)[e^{-ip\cdot z} + e^{ip\cdot z}] \sum_\lambda \hat{\mathbf{z}} \cdot \varepsilon(p, \lambda) \qquad (11.46)$$
$$\equiv \hat{\mathbf{z}}^i \, Y_{BB}(t, z)$$

where $\hat{\mathbf{z}} = \vec{\mathbf{z}}/|\vec{\mathbf{z}}|$ is the unit 3-vector in the direction of $\vec{\mathbf{z}}$, $z = |\vec{\mathbf{z}}|$, and p·z = $\omega(t - z \cos\theta)$. We define a spatial coordinate system – choosing the z-axis parallel to $\vec{z}$. Then we have

$$\vec{z} = (0, 0, z) \tag{11.47}$$
$$\vec{p} = (\sin\theta\cos\phi , \sin\theta\sin\phi, \cos\theta) \tag{11.48}$$
$$\vec{\varepsilon}(p,1) = (\cos\theta\cos\phi , \cos\theta\sin\phi, -\sin\theta) \tag{11.49}$$
$$\vec{\varepsilon}(p,2) = (-\sin\phi , \cos\phi, 0) \tag{11.50}$$

with the result (taking account of eq. 11.46)

$$Y_{BB}(t, z) = <BB, T| \, \hat{\mathbf{z}}\cdot \mathbf{Y}(t, z) \,|BB, T>$$
$$= 2\pi\int_{0}^{\infty} d\omega \; \omega^2 N_0(\omega) f(\omega,T) \int_{0}^{\pi} d\theta \; \sin^2\theta \; [e^{-ip\cdot z} + e^{ip\cdot z}] \tag{11.51}$$

where p·z = $\omega(t - z \cos\theta)$ with z = $|\vec{z}|$. We will develop integral representations and approximations to Y_{BB} in a later section.

11.4.4 Expectation Values of the Scale Factor A(t, X) and the Invariant Interval $d\tau^2$

The scale factor

$$b_M(X) = :2aM_c k^{-\frac{1}{2}}[1 + \alpha^2 M_c^2 X^2]^{-1}: \tag{11.52}$$

where α is a constant given by eq. 10.6.3, is a normal-ordered q-number expression. We can formally expand (define) this expression as a power series of normal-ordered powers of X^2 and then evaluate it between blackbody coherent states. First we note that

$$<BB, T|:Y^{i_1}(z)Y^{i_2}(z)Y^{i_3}(z)Y^{i_4}(z) \ldots Y^{i_n}(z):|BB, T> = \hat{z}^{i_1}\hat{z}^{i_2}\hat{z}^{i_3}\hat{z}^{i_4} \ldots \hat{z}^{i_n}(Y_{BB}(t, z))^n \tag{11.53}$$

We now set $z^\mu = y^\mu$, and use Y_{BB} to represent $Y_{BB}(t, \mathbf{y})$:

$$Y_{BB} \equiv Y_{BB}(t, \vec{y}) \tag{11.54}$$

Thus

$$b_{BB}(y, t) = <BB, T|b_M(X)|BB, T>$$

$$= 2aM_c k^{-\frac{1}{2}}\{1 + a^2[M_c^2 y^2 + 2i\, y Y_{BB} - Y_{BB}^2/M_c^2]\}^{-1} \tag{11.55}$$

and

$$B_{BB}(t, y) = <BB, T|B(t, X)|BB, T> = a(t)b_{BB}(y, t) \tag{11.56}$$

by eq. 11.29-31 where α is a constant (eq. 10.6.3). The expectation value of the q-number invariant interval (eq. 11.29) is the c-number expression:

$$d\tau_{BB}^2 = <BB, T|d\tau^2|BB, T>$$

$$= dt^2 - B_{BB}^2(t, y)[dX_{BB}^2 + X_{BB}^2(d\theta^2 + \sin^2\theta\, d\varphi^2)] \tag{11.57}$$

where

$$X_{BB} = y + iY_{BB}/M_c^2 \tag{11.58}$$

and

$$dX_{BB} = dy(1 + iM_c^{-2}\partial Y_{BB}/\partial y) \tag{11.59}$$

The appearance of Y_{BB} in the expression for the invariant interval (eq. 11.57) has two effects: it introduces complex space-time into the model and the Extended Robertson-Walker metric is no longer equivalent to the Robertson-Walker metric as it would be in the classical case.

We will see that Y_{BB} approaches zero at large times thus leading to the conventional Robertson-Walker models. But at small times of the order of the Planck time near the Big Bang we enter a brave new world of complex space-time. We will investigate the nature of this new complex world in the succeeding sections.

11.4.5 Representation and Approximations for $Y_{BB}(t, z)$

The angle integral in eq. 11.51 can be performed to yield

$$Y_{BB}(t, z) = \pi^{\frac{1}{2}}z^{-1} \int_0^{\infty} d\omega \, \omega^{-1}(e^{\omega/\kappa T} - 1)^{-\frac{1}{2}}\cos(\omega t)J_1(\omega z) \qquad (11.60)$$

where $J_1(\omega z)$ is a Bessel function using 3.915.5 of Gradshteyn (1965).

11.4.5.1 Some Representations of Y_{BB}

The integral in eq. 11.60 does not appear to be simply expressible in terms of standard transcendental functions. A series representation of the integral can be obtained by expanding the exponential factor due to the Planck distribution:

$$Y_{BB}(t,z) = \frac{1}{2}\pi^{\frac{1}{2}}\kappa T \sum_{n=0}^{\infty} (2n)![2^{2n}(n!)^2]^{-1}\{(2n+1+2i\kappa Tt)^{-1}F(\frac{1}{2}, 1; 2; -[\kappa Tz/(n+\frac{1}{2}+i\kappa Tt)]^2) +$$
$$+ (2n+1 - 2i\kappa Tt)^{-1} F(\frac{1}{2}, 1; 2; -[\kappa Tz/(n+\frac{1}{2} - i\kappa Tt)]^2)\} \qquad (11.61)$$

where $F(a, b; c; w)$ is a hypergeometric function.[70]
Using an integral representation[71]

$$F(a, b; c; w) = \Gamma(c)[\,\Gamma(b)\Gamma(c - b)]^{-1} \int_0^1 dt \, t^{b-1}(1 - t)^{c-b-1}(1 - tw)^{-a}$$

for $F(\frac{1}{2}, 1; 2; w)$ we see eq. 11.61 can be written in terms of simpler algebraic expressions:

$$Y_{BB}(t, z) = \frac{1}{2}\pi^{\frac{1}{2}}(\kappa Tz^2)^{-1} \sum_{n=0}^{\infty}(2n)![2^{2n}(n!)^2]^{-1}\{[(n+ \frac{1}{2} + i\kappa Tt)^2 + (\kappa Tz)^2]^{\frac{1}{2}} +$$
$$+ [(n+ \frac{1}{2} - i\kappa Tt)^2 + (\kappa Tz)^2]^{\frac{1}{2}} - (2n+1)\} \qquad (11.62)$$

Eq. 11.62 shows the limit of $Y_{BB}(t, z)$ for large t is

---

[70] Based on the integral 6.613.1 on p. 711 of Gradshteyn (1965).
[71] Magnus (1949) p. 8.

$$Y_{BB}(t, z) \rightarrow \pi^{\frac{1}{2}}\kappa T \sum_{n=0}^{\infty}(2n)![2^{2n+2}(n!)^2]^{-1}(2n + 1)[(n + \frac{1}{2})^2 + (\kappa Tt)^2]^{-1} \rightarrow 0 \qquad (11.63)$$

if $tT \rightarrow \infty$ as $t \rightarrow \infty$ as we see in cosmological models (see section 11.4).
     Thus

$$(b_{BB}(y, t))^2 dX^2 \rightarrow 2aM_c^2k^{-1}[1 + a^2M_c^2y^2]^{-2}dy^2 \equiv 2ak^{-1}[1 + a^2\check{r}^2]^{-2}d\check{r}^2 \qquad (11.64)$$

where α is a constant (eq. 10.6.3), for large t showing the Two-Tier cosmological model becomes a Robertson-Walker model at large times. However, the Two-Tier standard cosmological model is very different at small times of the order of the Planck time near the Big Bang point.

11.4.5.2 Approximate Solution for Y_{BB}

     The integral representation and power series representation of $Y_{BB}(t, y)$ do not reveal the physical behavior of the model for small times t and distances y. Therefore we will examine an approximation for $Y_{BB}(t, y)$ for ranges of y, t and T that are relevant for our considerations. We begin by scaling the integration variable in eq. 11.60 with the result:

$$Y_{BB}(t, y) = \pi^{\frac{1}{2}}y^{-1}\int_0^{\infty} d\omega\ \omega^{-1}(e^{\omega} - 1)^{-\frac{1}{2}}\cos(\omega\kappa Tt)J_1(\omega\kappa Ty) \qquad (11.65)$$

The blackbody exponential factor $(e^{\omega} - 1)^{-\frac{1}{2}}$ in the integrand of $Y_{BB}(t, y)$ enables the leading order approximate behavior of $Y_{BB}(t, y)$ to be determined for $0 \leq t \lesssim 10^{108}$ s – for all time, practically speaking. In a later section (section 11.3.4) we will see that our approximation to the integral in eq. 11.65 is consistent with the solution that we obtain for Y_{BB}, for the scale factor and thus for the temperature T. The approximations that we will make in eq. 11.65 are

$$\cos(\omega\kappa Tt) \approx 1 \qquad (11.66)$$
$$J_1(\omega\kappa Ty) \approx \omega\kappa Ty/2 \qquad (11.67)$$

They are based on $\kappa Tt \ll 1$ and $\kappa Ty \ll 1$ for all y ($0 \le y \le M_c^{-1}$). The exponential factor tends to limit contributions to the integral to small ω. After making these approximations we find

$$Y_{BB}(t, y) \simeq \tfrac{1}{2}\, \pi^{\frac{1}{2}}\kappa T \int_0^\infty d\omega\, (e^\omega - 1)^{-\frac{1}{2}}$$

$$\simeq \pi^{3/2}\kappa T/2 \tag{11.68}$$

The limit as t gets large can also be approximately determined from eq. 11.65. For large t such that κTy is small (and ω is small due to the exponential Planck distribution factor) we can again approximate the Bessel function with its leading power series expansion term and the exponential factor can again be approximated by $e^\omega - 1 \approx \omega$ so that eq. 11.65 becomes approximately

As $t \to \infty$:
$$Y_{BB}(t, y) \simeq \pi^{\frac{1}{2}}2^{-1}\, \kappa T \int_0^\infty d\omega\, \omega^{-\frac{1}{2}}\cos(\omega\kappa Tt)$$

$$= \pi 2^{-3/2}[\kappa T/t]^{\frac{1}{2}} \tag{11.69}$$

using 3.751.2 of Gradshteyn (1965). We note that, while κTy is small, κTt could possibly have been large in either a matter-dominated or radiation-dominated universe since it grows as t to a positive power (see section 11.3.) *However, since $Y_{BB}(t, y)$ approaches zero for large times its impact can only be seen in the initial formative stages of the universe near t = 0.*

11.5 The Scale Factor a(t) Near t = 0

The "time factor" a(t) of the scale factor $B_{BB}(t, y)$ appears in

$$B_{BB}(t, y) = <BB, T|B(t, X)|BB, T> = a(t)b_{BB}(y, t) \tag{11.70}$$

and is determined by the classical Einstein equation:

$$\dot{a}^2(t) - 8\pi G\rho_{tot}a^2(t)/3 = -k \tag{11.71}$$

As we have argued earlier, the source determining a(t) for small times in the neighborhood of t = 0 (the time of the Big Bang) is a large, macroscopic, classical density $\rho_{tot}(t)$ and thus a(t) may be considered to be a c-number quantity determined by the c-number Einstein equation to good approximation. This approximation should continue to hold even if this macroscopic density becomes enormous as t → 0. The quantum effects near t = 0 in the Two-Tier model, that we have developed, appear in the factor $b_{BB}(y, t)$, which we evaluated in previous sections.

11.6 A Complex Blackbody Temperature Near t = 0

The blackbody temperature T for relativistic particles (presumably the dominant type of particles near t = 0) is inversely proportional to the scale factor. <u>At large times</u> the blackbody temperature has the form

$$T = T_0/a(t) \tag{11.72}$$

where T_0 is a constant.

At time periods in the neighborhood of t = 0 (the Big Bang) space has three complex dimensions in the Two-Tier model. Temperature can be viewed as a measure of the root mean square speed (or the "average energy") of the components of the perfect fluid that we have assumed. In the case of a gas of particles of average energy E:

$$T = E/(3k/2) \tag{11.73}$$

In a complex space it is quite natural for the root mean squared speed to be complex as well. As a result complex temperatures naturally follow. Thus we will define

$$T = T_0/B_{BB}(t, y) \tag{11.74}$$

<u>for all time since t ≥ 0</u>. Since $B_{BB}(t, y)$ approaches $M_c a(t)b(\hat{r})$ at large times we find its large time behavior is consistent with those of standard Robertson-Walker models. At times near t = 0, the blackbody temperature T is complex since space is complex and

complex kinetic energy is allowed. Then we apply a Complex Lorentz (Reality) group transformation to obtain the physical temperature.

In the case of the complex temperature T above, the corresponding physical temperature is its absolute value (obtained by multiplying T by a phase factor from the Reality group)

$$T_{physical} = |T_0/B_{BB}(t, y)| \qquad (11.74a)$$

We shall use eq. 11.74 for the temperature, transforming the results below, afterwards, to real values using the 4-dimensional Reality group.

11.7 The Nature of the Universe Near t = 0

At this point we are ready to examine the Two-Tier model for the Big Bang period.

11.7.1 Behavior of the Complete Scale Factor B(t, y) Near t = 0

The behavior of the expectation value of the scale factor B(t, y), under the assumption that the Y quanta have a blackbody spectrum, is described by the equations:

$$b_{BB}(y, t) = 2aM_c k^{-\frac{1}{2}}\{1 + a^2[M_c^2 y^2 + 2i\, y Y_{BB} - Y_{BB}^2/M_c^2]\}^{-1} \qquad (11.75)$$
$$B_{BB}(t, y) = <BB, T|B(t, X)|BB, T> = a(t)b_{BB}(y, t) \qquad (11.76)$$

where α is a constant (eq. 10.6.3), and

$$Y_{BB}(t, y) \simeq \pi^{3/2}\kappa T/2 \qquad (11.77)$$
$$a(t) = [2\pi G\rho_0 n^2/3]^{1/n}\, t^{2/n} \qquad (11.78)$$
$$T = T_0/B_{BB}(t, y) \qquad (11.79)$$

as t → 0. *We will set the constant α = 1* in the interests of simplicity knowing that this value results in a metric fully equivalent to the Robertson-Walker metric at large times.

(Other values of a would also result in a metric equivalent to the Robertson-Walker metric at large times after a re-scaling of the radial coordinate.) Thus we may write

$$B_{BB}(t, y) \cong 2k^{-\frac{1}{2}}M_c a(t)\{1 + M_c^2 y^2 + iy\pi^{3/2}\kappa T_0/B_{BB} - \pi^3\kappa^2 T_0^2/(4M_c^2 B_{BB}^2)\}^{-1} \quad (11.80)$$

This quadratic algebraic equation for B_{BB} has the solutions:

$$B_{BB}(t, y) \cong (1 + M_c^2 y^2)^{-1}\{-i\varkappa M_c y + k^{-\frac{1}{2}}M_c a(t) \pm [\varkappa^2 - 2i\varkappa y k^{-\frac{1}{2}}M_c^2 a(t) + k^{-1}M_c^2 a^2(t)]^{\frac{1}{2}}\}$$
$$(11.81)$$

with

$$\varkappa = \pi^{3/2}\kappa T_0/(2M_c) \quad (11.82)$$

As t gets very large we obtain the equivalent of the Robertson-Walker metric scale factor in this approximation if we choose the plus sign in eq. 11.81 (assuming $a^2(t)$ becomes very large so other terms within the square root can be neglected):

$$B_{BB}(t, y) \rightarrow 2k^{-\frac{1}{2}}M_c a(t)/(1 + M_c^2 y^2) \quad (11.83)$$

Thus we must choose the plus sign in eq. 11.81:

$$B_{BB}(t, y) \cong (1 + M_c^2 y^2)^{-1}\{-i\varkappa M_c y + k^{-\frac{1}{2}}M_c a(t) + [\varkappa^2 - 2i\varkappa y k^{-\frac{1}{2}}M_c^2 a(t) + k^{-1}M_c^2 a^2(t)]^{\frac{1}{2}}\}$$
$$(11.84)$$

At t = 0 (the Big Bang) eq. 11.84 simplifies to (assuming a(0) = 0)

$$B_{BB}(0, y) \cong [\varkappa - i\varkappa M_c y]/(1 + M_c^2 y^2) \quad (11.85)$$

For small y, the real part of $B_{BB}(0, y)$ is a constant and the imaginary part of $B_{BB}(0, y)$ is proportional to y.

11.7.2 The Expectation Value of the Scale Factor $A_{BB}(0, y)$ near $t = 0$

Eq. 11.85 gives the approximate behavior of the expectation value of the scale factor $B_{BB}(0, y)$ near $t = 0$ as a function of y. If we compare eq. 11.85 with the mechanism described in eqns. 11.1.6 – 11.1.9 of section 11.1 for cancelling the a(t) factor within the complete scale factor we see that we have found the blackbody spectrum of the Y quanta implements this mechanism. The solution can be written in the form:

$$A_{BB}(t, y) = M_c^{-1}B_{BB}(t, y) \cong \beta_0(y) + \beta_1(y)a(t) + \ldots \qquad (11.86)$$
$$\beta_0(y) = x(1 - iM_c y)/[M_c(1 + M_c^2 y^2)] \qquad (11.87)$$
$$\beta_1(y) = k^{-\frac{1}{2}}(1 - iM_c y)/(1 + M_c^2 y^2) \qquad (11.88)$$

Eqns. 11.86–11.88 are expressed in terms of the y variable. They can be expressed in terms of ř as:

$$A(t, ř) = a(t)b(ř) = 2\alpha k^{-\frac{1}{2}}a(t)[1 + \alpha^2 ř^2]^{-1} \qquad (11.89)$$

where α is set to $a = 1$. Evidently, we have $M_c y \equiv ř$ at the level of approximation that we are using. Furthermore we can use

$$ř = \{[1 - (1 - kr^2)^{\frac{1}{2}}]/[1 + (1 - kr^2)^{\frac{1}{2}}]\}^{\frac{1}{2}} \qquad (11.90)$$

to express ř in terms of the Roberson-Walker radial coordinate r. Thus we find that the Robertson-Walker scale factor a(t) becomes

$$a(t) \rightarrow (1 + M_c^2 y^2)(2k^{-\frac{1}{2}})^{-1}A_{BB}(t, y) \equiv a_{BBRW}(t, ř) \qquad (11.91)$$
$$a_{BBRW}(t, ř) \cong \beta_{0RW}(ř) + \beta_{1RW}(ř)a(t) + \ldots \qquad (11.92)$$

using the subscript "RW" to denote quantities scaled to the standard Robertson-Walker metric, with

$$\beta_{0RW}(ř) = x(1 - iř)/[2k^{-\frac{1}{2}}M_c] \qquad (11.93)$$

$$\beta_{1RW}(\check{r}) = (1 - i\check{r})/2 \qquad\qquad (11.94)$$

where $\check{r}$ is specified by eq. 11.90.

Using the Reality group we can "rotate" the complex scale factor, which is a factor in coordinate expressions, to a real value—its absolute value

$$a(t) \rightarrow (1 + M_c^2 y^2)(2k^{-\frac{1}{2}})^{-1}|A_{BB}(t, y)| \equiv |a_{BBRW}(t, \check{r})| \qquad (11.91a)$$

We will study the implications of this scale factor in the following chapters.

12. Big Bang Scale Factor

The previous chapters have examined the quantum part of the scale factor of the Quantum Extended Robertson-Walker Model. We found that the quantum scale factor had the form

$$A(t, \check{r}) = a(t)b(\check{r}, t) \tag{12.1}$$

where $a(t)$ is a c-number scale factor dependent on the total mass-energy density of the universe, and $b(\check{r}, t)$ is a quantum scale factor dependent primarily on the second quantized Y^μ field (the imaginary part of the quantum coordinate X^μ.)

The expectation value of $A(t, \check{r})$ had the form

$$<A(t, \check{r})> = a(t)<b(\check{r}, t)> \tag{11.7}$$

We evaluated the $<b(\check{r}, t)>$ in chapter 10 under the assumption that Y quanta had a black body energy distribution. We continue the discussion in chapter 13.

13. From t = 0 to t = t_T: a(t) Scale Factor and Hubble Constant

13.1 Introduction

This chapter develops a numerical model for the scale factor of the Big Bang metastable state. Although the universe that we live in is almost flat according to recent WMAP experimental data, the small curvature of space closes the universe and has significant effects. Therefore we will not use a flat space approximation. Our goal is to obtain an order of magnitude understanding of the evolution of the universe from the beginning. Our calculated numerical quantities appear generally to be of the right order of magnitude. And the physical ideas appear to be consistent with a reasonable view of reality.

Chapter 11 described the physical implications of quantum part b($\check{r}$, t) of the total scale factor A($\check{r}$, t). It depends on the black body Y quanta.

13.1 Total Mass-Energy Density from t = 0 to t = t_T

In this chapter we consider a(t), which we take to be a c-number quantity since it depends solely on macroscopic mass-energy densities as seen in

$$\rho_{tot}(t) = \rho_{crit}[\Omega_\gamma(t) + \Omega_\Lambda(t) + \Omega_m(t) + \Omega_d(t)] \equiv \rho_{crit}\Omega_H(t) \qquad (9.7)$$

The data that we use in this chapter and throughout are the combined results of the WMAP and SDSS data[72] based on the assumption of a non-flat space. In particular we use the following values:

$$h = \text{Hubble parameter} = 0.678(9)$$

---

[72] M. Tanabashi *et al* (Particle Data Group), Phys. Rev. D**98**, 030001 (2018).

$$\rho_{crit} = \text{Critical density} = 1.87840(9) \, h^2 \times 10^{-29} \, \text{g/cm}^{-3}$$

$$\Omega_\Lambda = \text{Dark Energy density}/\rho_{cr} = \rho_{de}/\rho_{cr} = 0.692 \pm 0.012$$

$$\Omega_d = \Omega_c = \text{Cold Dark matter density}/\rho_{cr} = \rho_c/\rho_{cr} = 0.1186(20) \, h^{-2}$$

$$\Omega_b = \text{Baryon density}/\rho_{cr} = \rho_b/\rho_{cr} = 0.02226 \, h^{-2}$$

$$\Omega_m = \Omega_b + \Omega_d$$

$$\quad = \text{pressureless Matter density}/\rho_{cr} = \rho_m/\rho_{cr} = 0.308 \pm 0.012$$

$$\Omega_{tot} = \Omega_m + \Omega_\Lambda = 1.000 \pm 0.024$$

$$t_0 = t_{now} = \text{Age of universe} = 13.80 \pm 0.04 \, \text{Gyr}$$

$$(13.1.1)$$

The time dependence of the scale parameter is given by the Einstein equation:

$$\dot{a}^2 - 8\pi G \rho_{tot} a^2/3 = -k \qquad (13.1.2)$$

. Also, based on an analysis of WMAP[73] data:

$$r_{universe}(t_{now}) = \text{visible radius of the universe} = 4.314 \times 10^{28} \, \text{cm} \qquad (13.1.3)$$

The reader is directed to the original papers for other parameters, error bars and a detailed analysis of the data. *We use units where $\hbar = c = 1$ unless stated otherwise.*
We also use:[74]

$$\Omega_\gamma = \text{radiation density}/\rho_{cr} = \rho_\gamma/\rho_{cr} = 2.473 h^{-2} \times 10^{-5} \, (T/2.7255)^4 h^{-2}$$

$$= \quad 5.38 \times 10^{-5} \qquad (13.1.4)$$

13.2 The Behavior of the Scale Factor a(t) after the Big Bang Period

The universe, as we know it today, contains a variety of forms of energy. The current densities of these forms of energy are listed in section 13.1. From them we can develop the form of the total energy density and project it back to the instants after the Big

---

[73] M. Tanabashi *et al* (Particle Data Group), Phys. Rev. D**98**, 030001 (2018).
[74] M. Tanabashi *et al op. cit.*

Bang. The Big Bang period is significantly different as we have seen in the preceding chapter.

The total mass-energy density as a function of time from eq. 9.6 neglecting $\Omega_T(t)$ is

$$\rho_{tot}(t) \equiv \rho_{crit}\Omega_{tot}(t) = \rho_{crit}[\Omega_\gamma(t) + \Omega_M(t) + \Omega_\Lambda(t)]$$

$$= \rho_{crit}[\Omega_\Gamma/a^4(t) + \Omega_M/a^3(t) + \Omega_\Lambda] \tag{13.2.1}$$

where we treat the time dependence of Dark energy as similar to "normal" energy, and of Dark Matter as similar to "normal" matter (as in the Unified SuperStandard Theory.) based on well-known arguments.[75] Note

$$\Omega_M(t) = \Omega_m(t) = \Omega_b(t) + \Omega_d(t)$$

The time dependence of the scale parameter is given by the Einstein equation:

$$\dot{a}^2 - 8\pi G\rho_{tot}a^2/3 = -k \tag{10.5.2}$$

where a(t) is the Robertson-Walker scale factor with $a(t_{now}) = 1$.
Before proceeding to the solution of eq. 10.5.2 we need to obtain a reasonable estimate of the curvature constant k. We can use the Robertson-Walker expression for the radius of the universe

$$r_{universe}(t_{now}) = a(t_{now})/k^{1/2} \tag{13.2.2}$$

or

$$k^{-1/2} = r_{universe}(t_{now}) \tag{13.2.3}$$

Using this radius we obtain one estimate

$$k = 5.37 \times 10^{-58}\,cm^{-2} \tag{13.2.4}$$

---

[75] Weinberg(1972), Dodelson(2003). We differ by not distinguishing Dark from normal based on our SuperStandard Theory.

13.2.1 General Form of a(t) Scale Factor Einstein Equation

We find the general form of the scale factor differential equation by combining eqs. 10.5.2 and 13.2.1

$$\dot{a}^2 - H_0^2 a^2(t)[\Omega_\Gamma/a^4(t) + \Omega_M/a^3(t)] = -k \tag{13.2.1.1}$$

where the Hubble constant H_0 satisfies

$$H_0^2 = [d(\ln a)/dt]|_{t \,=\, t_{\text{now}}} = 8\pi G\rho_{cr}/3 \equiv 1.2 \times 10^{-56} h^2 \text{ cm}^{-2} \tag{13.2.1.2a}$$

or

$$H_0 = 100 \text{ h km s}^{-1} \text{ Mpc}^{-1} = h \times (9.777752 \text{ Gyr})^{-1} =$$
$$= 1.1 \times 10^{-28} h \text{ cm}^{-1} \equiv 3.24 \times 10^{-18} h \text{ s}^{-1} \tag{13.2.1.2b}$$

If we evaluate eq. 13.2.1.1 for the present time we find

$$k = [\Omega_\gamma + \Omega_\Lambda + \Omega_m]H_0^2 \cong [\Omega_m + \Omega_\Lambda]H_0^2 = H_0^2$$

$$= 5.56 \times 10^{-57} \text{ cm}^{-2} \tag{13.2.1.2c}$$

since

$$\Omega_m + \Omega_\Lambda \cong 1.00$$

We now define

$$\xi = k/H_0^2 \cong 1 \tag{13.2.1.3}$$

for later use.

The solution of eq. 13.2.1.1 can be put into the form of integrals representing combinations of elliptic integrals:

$$\int_{a(t')}^{a(t)} da \, a \, H_0^{-1} \left[\Omega_\Lambda a^4 - \xi a^2 + \Omega_m a + \Omega_\gamma\right]^{-\frac{1}{2}} = \int_{t'}^{t} dt \tag{13.2.1.4}$$

The result of these integrations is an implicit equation for a(t) that cannot be expressed in a simple closed form in terms of known functions. This equation can be easily solved numerically. A graph of a(t) is displayed in Fig. 13.2.1.1. Although it looks linear there are significant non-linearities in various parts of the plot of a(t). The radiation-dominated phase is not visible. It is a small slice of the plot since it amounts to less than 10^{13} s.

Because we know physically that approximations are possible for each of the various epochs: the matter-dominated epoch, the radiation-dominated epoch and so on, we can find physically meaningful approximations for each epoch. We therefore provisionally divide the life of the universe into two epochs: an explosive growth epoch, and an expanding epoch subdivided into matter-dominated and radiation-dominated phases.

We begin with the most recent phase, the explosive growth epoch, and then describe the previous phase, the expanding epoch, containing the matter dominated and radiation dominated phases. The results below in this chapter assume that $T > T_c$ implying no influx of Megaverse mass-energy. Chapter 14 considers the impact of a Megaverse mass-energy influx.

13.2.2 The Explosive Growth Epoch

The integral on the left in eq. 13.2.1.4 appears to support a simple approximation during the explosive growth phase. Notice that at $t = t_{now}$ we have $-\xi a^2 + \Omega_m a + \Omega_\gamma \cong -0.602$. The sum of these terms gets smaller as we proceed into the past. Therefore we approximate the left side with

$$\int_{a(t')}^{a(t)} da \, aH_0^{-1}[\Omega_\Lambda a^4]^{-\frac{1}{2}} = (H_0\Omega_\Lambda^{\frac{1}{2}})^{-1} \int_{a(t')}^{a(t)} da/a = (H_0\Omega_\Lambda^{\frac{1}{2}})^{-1}\ln[a(t)/a(t')] \cong t - t' \qquad (13.2.2.1)$$

or

$$a(t) = a(t')\exp[H_0\Omega_\Lambda^{\frac{1}{2}}(t - t')] \qquad (13.2.2.2)$$

Thus we have deSitter-like exponential growth.[76] We would expect this growth phase to last approximately for the period where

---

[76] The Explosive Growth Epoch is also known as the *Steady State Cosmology*. It was pioneered by F. Hoyle, M.N.R.A.S. **108**, 372 (1948); F. Hoyle and J. V. Narlikar, Proc. R. Soc. London **A290**, 162 (1966); H. Bondi and T. Gold, M.N.R.A.S. **108**, 252 (1948) and references therein. It implements a "perfect cosmological principle" by

$$\Omega_\Lambda a^4(t_E) \approx \Omega_m a(t_E) \implies a(t_E) = .76 \qquad (13.2.2.3)$$

until the present where t_E is the time when $a(t_E) = .76$. Since we have normalized $a(t)$ by

$$a(t_{now}) = 1 \qquad (13.2.2.4)$$

we see that eq. 13.2.2.2 requires

$$a(t) = \exp[H_0\Omega_\Lambda^{\frac{1}{2}}(t - t_{now})] \qquad (13.2.2.5)$$

in the epoch that we have called the Explosive Growth Epoch. Note

$$H_0\Omega_\Lambda^{\frac{1}{2}} = 1.83\times 10^{-18} \text{ s}^{-1} \qquad (13.2.2.6)$$

---

giving a universe that is the same at all points and times. It requires the continuous creation of matter with the origin of the created matter uncertain although it will be an enormous amount since it is distributed over the universe. In chapter 14 we consider a more detailed theory in which "new" matter originates in the external Megaverse as a result of universe surface tension.

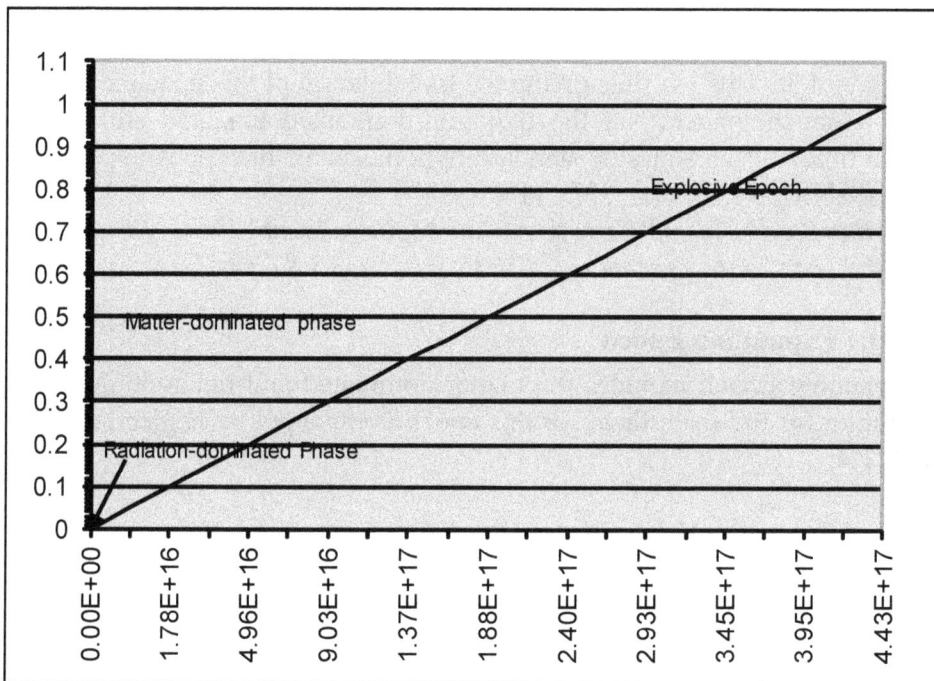

Figure 13.2.1.1. A plot of a(t) generated from eq. 13.2.1.4 through numerical integration with time measured in seconds.

Epoch	Type	Phases
I	**Explosive Growth**	
II	**Expanding**	**Matter-Dominated** **Radiation-Dominated**

Table 13.2.1.1. Epochs and phases of the universe after the Big Bang Epoch.

The beginning of this epoch is set by eq. 13.2.2.3:

$$t_E = t_{now} + (H_0\Omega_\Lambda^{1/2})^{-1}\ln(.76) = 2.87 \times 10^{17} \text{ s} \qquad (13.2.2.7)$$

where $t_{now} = 4.35 \times 10^{17}$ s , thus giving the time interval of this epoch to be 1.47×10^{17} s = 4.7 Gyr —much longer than the radiation-dominated era, and still expanding. The transition time t_E is close to the standard hypothesis for the appearance of Dark Energy of at a red-shift $z \sim .5$ or $a(t) = 2/3$. $(1 + z = a^{-1})$

   Thus the Explosive Growth period begins around 2.87 $\times 10^{17}$ s (t_E = 9.1 Gyr), extends 4.7 Gyrs to the present t_{now} = 4.35 $\times 10^{17}$ s (13.8 Gyrs).

13.2.3 The Expanding Epoch

The Expanding Epoch includes the matter-dominated and radiation-dominated phases. The solution for the scale factor in this epoch is obtained by neglecting the Ω_Λ term in eq. 13.2.1.4:

$$\int_{a(t')}^{a(t)} da\, a\, H_0^{-1}\, [-\xi a^2 + \Omega_m a + \Omega_\gamma]^{-1/2} = \int_{t'}^{t} dt \qquad (13.2.3.1)$$

This equation is easily integrated yielding:

$$-\xi H_0(t - t') = [-\xi a^2(t) + \Omega_m a(t) + \Omega_\gamma]^{1/2} + \Omega_m(2\xi^{1/2})^{-1}\arcsin[(\Omega_m - 2\xi a(t))(\Omega_m^2 + 4\xi\Omega_\gamma)^{-1/2}] -$$

$$- [-\xi a^2(t') + \Omega_m a(t') + \Omega_\gamma]^{1/2} - \Omega_m(2\xi^{1/2})^{-1}\arcsin[(\Omega_m - 2\xi a(t'))(\Omega_m^2 + 4\xi\Omega_\gamma)^{-1/2}]$$

$$(13.2.3.2)$$

Letting t' = 0 assuming a(t' = 0) = 0 and we find

$$-\xi H_0 t = [-\xi a^2(t) + \Omega_m a(t) + \Omega_\gamma]^{1/2} + \Omega_m(2\xi^{1/2})^{-1}\arcsin[(\Omega_m - 2\xi a(t))(\Omega_m^2 + 4\xi\Omega_\gamma)^{-1/2}] -$$

$$- \Omega_\gamma^{1/2} - \Omega_m(2\xi^{1/2})^{-1}\arcsin[\Omega_m(\Omega_m^2 + 4\xi\Omega_\gamma)^{-1/2}] \qquad (13.2.3.3)$$

Eq. 13.2.3.3 can be substantially simplified. Note the arguments of the arcsines are both near one in value due to the smallness of Ω_γ and ξ. Both arcsines can be approximated using

$$\arcsin(\,1-\epsilon) \cong \pi/2 - (2\epsilon)^{\frac{1}{2}} \qquad (13.2.3.4)$$

for small ϵ.

First we approximate the arguments of the arcsines with

$$\arcsin[(\Omega_m - 2\xi a(t))(\,\Omega_m^2 + 4\xi\Omega_\gamma)^{-\frac{1}{2}}] \cong \arcsin(1 - 2\xi a(t)\Omega_m^{-1} - 2\xi\Omega_\gamma\Omega_m^{-2})$$

and

$$\arcsin[\Omega_m(\,\Omega_m^2 + 4\xi\Omega_\gamma)^{-\frac{1}{2}}] \cong \arcsin(1 - 2\xi\Omega_\gamma\Omega_m^{-2})$$

Then using eq. 13.2.3.4 we obtain

$$[-\xi a^2(t) + \Omega_m a(t) + \Omega_\gamma]^{\frac{1}{2}} - [\Omega_m a(t) + \Omega_\gamma]^{\frac{1}{2}} \cong -\xi H_0 t \qquad (13.2.3.5)$$

Noting that $\xi a^2(t)$ is much smaller than the other terms in the first square root in eq. 13.2.3.5 we can further approximate that equation further by expanding the square root to obtain:

$$a^2(t)[\Omega_m a(t) + \Omega_\gamma]^{-\frac{1}{2}} \cong 2H_0 t \qquad (13.2.3.6)$$

Eq. 13.2.3.6 embodies the standard matter-dominated and radiation-dominated expressions for the scale factor.

Matter-Dominated Phase

In the case of the matter-dominated phase we have

$$\Omega_m a(t) > \Omega_\gamma$$

We can approximate eq. 13.2.3.6 accordingly

$$a^2(t)[\Omega_m a(t)]^{-\frac{1}{2}} \cong 2H_0 t \qquad (13.2.3.7)$$
$$a(t) \cong [2H_0\Omega_m^{\frac{1}{2}}t]^{2/3} \qquad (13.2.3.8)$$

Radiation-Dominated Phase

In the case of the radiation-dominated phase we have

$$\Omega_m a(t) < \Omega_\gamma$$

due to the smallness of the scale factor in that phase We approximate eq. 13.2.3.6 accordingly

$$a^2(t)[\Omega_\gamma]^{-\frac{1}{2}} \cong 2H_0 t \tag{13.2.3.9}$$

Thus

$$a(t) \cong [2H_0\Omega_\gamma^{\frac{1}{2}}t]^{\frac{1}{2}} \tag{13.2.3.10}$$

The crossover point between the radiation-dominated and matter-dominated phase is at

$$\Omega_m a(t_{RM}) = \Omega_\gamma \qquad \text{or} \qquad a(t_{RM}) = 1.8 \times 10^{-4} \tag{13.2.3.11}$$

where t_{RM} is the crossover time:

$$t_{RM} = 7 \times 10^{11} \text{ s} = 2.22 \times 10^{-5} \text{ Gyr} \tag{13.2.3.12}$$

which indicates a very short radiation-dominated phase. Figs. 13.2.3.1 and 13.2.3.2 contain plots of a(t). Fig. 13.2.3.1 shows the approximate implicit equation for a(t) (eq. 13.2.3.6) is quite good over the entire range, and particularly good for small times in the radiation-dominated time frame. Since this region is the region of interest as it connects to the Big Bang Epoch we shall use this approximation, and eq. 13.2.3.10, for a(t) at very small times near t = 0.

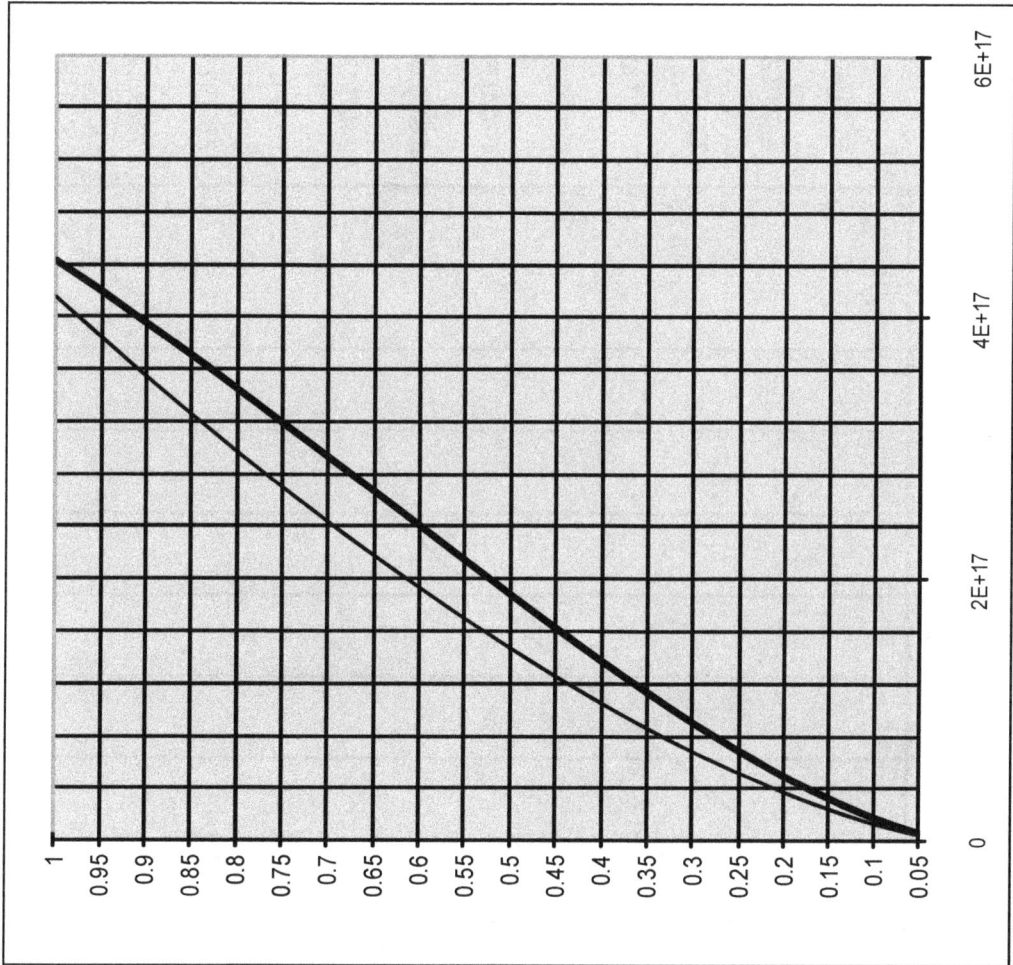

Figure 13.2.3.1. A plot of a(t) (horizontal axis) vs. time in seconds. The thick line is the plot of a(t) obtained by direct numerical integration of eq. 13.2.1.4 including the three density terms and the curvature constant term. The thin line is a plot of a(t) calculated directly from the approximation eq. 13.2.3.6.

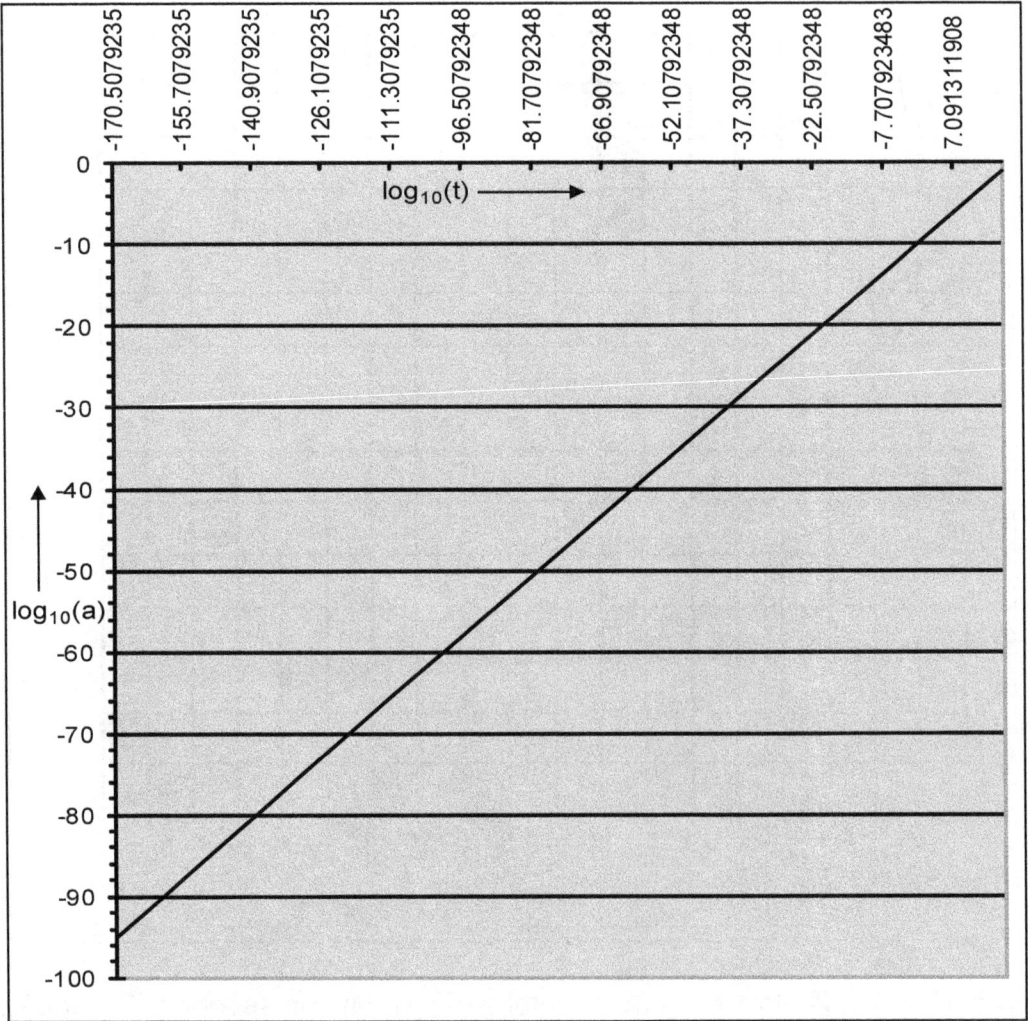

Figure 13.2.3.2. A log-log (base 10) plot of a(t) vs. t (in seconds) for small times calculated from eq. 13.2.3.6.

13.3 Two-Tier Quantum Big Bang Model at the Beginning of Time

Section 13.2 presented the approximate expressions for the scale factor near t = 0. In this section we will consider numerical estimates for the scale factor, and other quantities of interest, such as the temperature and density near, and at, t = 0 in the Two-Tier Hybrid Big_Bang model whose early time behavior is described in chapter 11.

In estimating quantities we are confronted with an imprecise determination of the needed input parameters because of experimental uncertainties and the impossibility of currently finding certain parameters experimentally in a model independent way. So we will use reasonable estimates for input parameters realizing that they may sometimes be off by up to a few orders of magnitude. Because of the vast differences in value between terms comprising the scale factors we believe a few orders of magnitude is often not a significant issue in determining the relative importance of terms.

The reader may notice slight differences in values due to rounding off numbers to three figures in the text while keeping values to 16 significant digits in the calculations. These differences can have a cumulative effect so we want to emphasize that our goal is order of magnitude accuracy.

The input data items are those of section 13.2 plus:

1. The current temperature of the Cosmic Microwave Background (CMB) radiation = 2.7255 °K.

2. We assume $M_c = M_{Planck} = 1.22 \times 10^{28}$ eV since M_{Planck} is the only large mass intrinsic to the theory of gravitation and thus seemed to be a natural choice.

13.3.1 The CMB Temperature

From the current CMB temperature (T = 2.7255 °K) we find

$$\kappa T = \kappa T_0/a(t_{now}) = \kappa T_0 = 2.35 \times 10^{-4} \text{ eV} \qquad (13.3.1.1)$$

For later use we define

$$x = \pi^{3/2}\kappa T_0/(2M_c) \cong 5.3 \times 10^{-32} \qquad (13.3.1.2)$$

13.3.2 The Generalized Robertson-Walker Scale Factor

A_{BB} is related to B_{BB} by eq. 11.86. Our approximation for the B_{BB} scale factor is:[77]

$$B_{BB}(t, y) \cong (1 + M_c^2 y^2)^{-1}\{-i\varkappa M_c y + k^{-\frac{1}{2}}M_c a(t) + [\varkappa^2 - 2i\varkappa yk^{-\frac{1}{2}}M_c^2 a(t) + k^{-1}M_c^2 a^2(t)]^{\frac{1}{2}}\}$$
$$(13.1.5)$$

If we let $y = M_c^{-1}$ then we can find B_{BB} "at the borders of the universe" where it simplifies to:

$$B_{BB}(t) = B_{BB}(t, M_c^{-1}) \cong \{-i\varkappa + \varpi a(t) + [\varkappa^2 - 2i\varkappa\varpi a(t) + \varpi^2 a^2(t)]^{\frac{1}{2}}\}/2 \quad (13.3.2.1)$$

with

$$\varpi = k^{-\frac{1}{2}}M_c = 8.31 \times 10^{60} \quad (13.3.2.2)$$

If t and a(t) are small, then

$$A_{BB} = A_{BB}(t, \check{r}) = M_c^{-1}B_{BB}(t, \check{r}) \cong \beta_0(\check{r}) + \beta_1(\check{r})a(t) +... \quad (13.3.2.3)$$

$$\beta_0(\check{r}) = \varkappa(1 - i\check{r})/[M_c(1 + \check{r}^2)] \quad (13.3.2.4)$$
$$\beta_1(\check{r}) = k^{-\frac{1}{2}}(1 - i\check{r})/(1 + \check{r}^2) \quad (13.3.2.5)$$

We noted earlier that if we transformed our results back to Robertson-Walker coordinates we would have a modified scale factor $a = a_{BBRW}$ which we can continue to express in terms of $\check{r} = M_c y$.

$$a(t) \rightarrow (1 + M_c^2 y^2)k^{\frac{1}{2}}A_{BB}(t, y)/2 \equiv a_{BBRW}(t, \check{r}) \quad (8.5.2.4)$$

Thus

$$a_{BBRW}(t, \check{r}) = (1 + M_c^2 y^2)(2k^{-\frac{1}{2}}M_c)^{-1}B_{BB}(t, y) \quad (13.3.2.6)$$

$$= \frac{1}{2}\{a(t) - i\varkappa\varpi^{-1}\check{r} + [(\varkappa/\varpi)^2 + a^2(t) - 2i(\varkappa/\varpi)\check{r}a(t)]^{\frac{1}{2}}\} \quad (13.3.2.7)$$

---

[77] Please note that the physical value of scale factors is the absolute value of the complex scale factors (obtained by use of the Reality group transformations of Blaha (2018e)) appearing here and in the following discussions.

by eq. 8.92. If we evaluate a_{BBRW} at $\check{r} = M_c y = 1$ (the maximum value of y), since it determines the "size" of the universe, then

$$a_{BBRW}(t) \equiv a_{BBRW}(t, 1) = \{a(t) - i\gamma + [\gamma^2 + a^2(t) - 2i\gamma a(t)]^{\frac{1}{2}}\}/2 \qquad (13.3.2.8)$$

with the dimensionless constant

$$\gamma = x/\varpi = 6.38 \times 10^{-93} \qquad (13.3.2.9)$$

This value reminds one of Eddington's famous remarks that cosmological quantities often have orders of magnitude that are approximately multiples of 90.

The real and imaginary parts of $a_{BBRW}(t)$ are:

$$\text{Re } a_{BBRW}(t) = a(t)/2 + [Q(t)(1 + \cos \psi(t))/2]^{\frac{1}{2}}/2 \qquad (13.3.2.10)$$

and

$$\text{Im } a_{BBRW}(t) = -\gamma/2 - [Q(t)(1 - \cos \psi(t))/2]^{\frac{1}{2}}/2 \qquad (13.3.2.11)$$

where

$$Q(t) = [(\gamma^2 + a^2(t))^2 + 4\gamma^2 a^2(t)]^{\frac{1}{2}} \qquad (13.3.2.12)$$

and

$$\cos \psi(t) = (\gamma^2 + a^2(t))/Q \qquad (13.3.2.13)$$

There are 2 distinctly different periods specified by the time dependence of a_{BBRW}. The first period corresponds to an initial universe of "slowly" increasing size. The second period is the radiation and matter dominated phases with a fairly rapid increase in a(t). The boundary time t_c between these periods is specified by a continuity condition using:

$$a(t) \cong [2\Omega_\gamma^{\frac{1}{2}} H_0 t]^{\frac{1}{2}} \qquad (13.2.3.10)$$

above t_c and

$$\text{Re } a_{BBRW}(t_c) = a(t_c)/2 + [Q(t_c)(1 + \cos \psi(t_c))/2]^{\frac{1}{2}}/2 = a(t_c)$$

below using eq. 13.2.3.10. Eq. 13.3.2.14 below is the self-consistency condition[78]

$$[2\Omega_\gamma^{1/2}H_0t]^{1/2} = a(t_c) \cong a_{BBRW}(t_c) = \gamma \qquad (13.3.2.14)$$

giving

$$t_c = 1.26 \times 10^{-165} \text{ s} \cong 10^{-165} \text{ s} \qquad (13.3.2.15)$$

In this period we find

$$\text{Re } a_{BBRW}(0) \cong \gamma/2 = 3.19 \times 10^{-93} \quad (13.3.2.16a)$$

$$\text{Re } a_{BBRW}(t_c) \cong 1.28\gamma = 8.166 \times 10^{-93} \qquad (13.3.2.16b)$$

$$\text{Im } a_{BBRW}(0) = -\gamma/2 = -3.19 \times 10^{-93} \qquad (13.3.2.16c)$$

with the radius, and volume, of the universe "slowly" increasing during this period. The nature of this period directly reflects the effects of the blackbody Y quanta. This can be seen from its inverse square dependence on M_c. As $M_c \to \infty$ the constant $\gamma \to 0$ and thus $a_{BBRW} \to 0$, with the universe scaling down to zero size with the attendant standard model catastrophes of infinite density and temperature., The blackbody quanta (Dark Energy) give the universe a short-lived meta-stable initial size thus avoiding catastrophic divergences.

---

[78] This continuity condition is based on the relation a(t) $a(t) \to a_{BBRW}(t, \v{r})$ in eq. 13.4.1 a(t) is embedded within $a_{BBRW}(t, \v{r})$.

Epoch	Type	Phases	Time Period
I	Explosive Growth	Dark Energy-dominated	2.87×10^{17} s $- 4.35 \times 10^{17}$ s
II	Expanding	Matter-dominated	7×10^{11} s $- 2.87 \times 10^{17}$ s
		Radiation-dominated	1.26×10^{-165} s $- 7 \times 10^{11}$ s
III	Metastable Big Bang	Blackbody Y quanta dominated	0 s $- 1.26 \times 10^{-165}$ s

Epoch	Type	Phases	Time Interval
I	Explosive Growth	Dark Energy-dominated	4.7 Gyr
II	Expanding	Matter-dominated	9.1 Gyr
		Radiation-dominated	2.22×10^{-5} Gyr
III	Metastable Big Bang	Blackbody Y quanta dominated	1.26×10^{-165} sec

Table 13.3.2.1 Epochs and phases of the Universe since t = 0.

The period after t_c is dominated by the usual scale factor a(t). This scale factor shows up directly in the real part of a_{BBRW}. Its behavior is:

$t < t_c$ (or a(t) < γ)

$$\text{Re } a_{BBRW}(t) \cong \gamma/2 + a(t)/2 \qquad (13.3.2.17a)$$

$t_c < t < t_{now}$

$$\text{Re } a_{BBRW}(t) \cong a(t) \qquad (13.3.2.17b)$$

Thus continuity (eq. 13.3.2.14) implies

t = t$_c$

$$\text{Re } a_{BBRW}(t_c) \cong \gamma = a(t_c) \qquad (13.3.2.17c)$$

The behavior of the imaginary part of a_{BBRW} is also indirectly dominated by a(t) but in a much less dramatic way. The gradual growth of the imaginary part of a_{BBRW} is more or less indicated by the following three values:

$$\text{Im } a_{BBRW}(0) \cong -\gamma/2 = -3.19 \times 10^{-93} \qquad (13.3.2.16c)$$
$$\text{Im } a_{BBRW}(t_c) \cong -0.822\gamma = -5.24 \times 10^{-93} \qquad (13.3.2.18a)$$
$$\lim_{t \to \infty} \text{Im } a_{BBRW}(t) = -\gamma = -6.38 \times 10^{-93} \qquad (13.3.2.18b)$$

and also the behavior

a(t) ≪ γ (or t < t$_c$)

$$\text{Im } a_{BBRW}(t) \cong -\gamma/2 - a(t)/2 \qquad (13.3.2.18c)$$

γ ≪ a(t) (or t$_c$ < t)

$$\text{Im } a_{BBRW}(t) \cong -\gamma + \mathcal{O}([\gamma/a(t)]^2) \cong -\gamma \qquad (13.3.2.18d)$$

Both the real and imaginary parts of a_{BBRW} roughly double in the time period [0, t$_c$] and thereafter we see a gradual increase of Im a_{BBRW} in absolute value from $-.5\gamma$ to $-\gamma$ over the lifetime of the universe.

After t$_c$ the real part grows dramatically (eq. 13.3.2.17) while the imaginary part remains minute. The details of the interpretation of the behavior of the scale factor a_{BBRW} in the Big Bang Epoch [0, t$_c$] will be explored in chapter 15. It suffices, for now, to say the universe in the period before t$_c$ is in a meta-stable state of "slowly" growing size due to the dynamics of the blackbody Y quanta. At t$_c$ the epoch of the expanding universe, as we know it, begins!

In differentiating between the real and imaginary parts of a_{BBRW} we must realize that the Reality group combines them into a single real quantity when scaling coordinates. Happily the small size of the imaginary part it can be neglected in most situations.

Before proceeding to describe physically interesting features of the early universe we note the radial dependence of the scale factor $a_{BBRW}(t, \v{r})$ in general:

$$a_{BBRW}(t, \v{r}) = \{-i\gamma\v{r} + a(t) + [\gamma^2 - 2i\gamma\v{r}a(t) + a^2(t)]^{\frac{1}{2}}\}/2 \quad (13.3.2.19)$$

$$\mathrm{Re}\ a_{BBRW}(t, \v{r}) = a(t)/2 + [R(t, \v{r})(1 + \cos\psi(t, \v{r})/2]^{\frac{1}{2}}/2 \quad (13.3.2.20)$$

$$\mathrm{Im}\ a_{BBRW}(t, \v{r}) = -\gamma\v{r}/2 - [R(t, \v{r})(1 - \cos\psi(t, \v{r}))/2]^{\frac{1}{2}}/2 \quad (13.3.2.21)$$

where

$$R(t, \v{r}) = [(\gamma^2 + a^2(t))^2 + 4(\gamma\v{r}a(t))^2]^{\frac{1}{2}} \quad (13.3.2.22)$$

and

$$\cos\psi(t, \v{r}) = (\gamma^2 + a^2(t))/R(t, \v{r}) \quad (13.3.2.23)$$

We find that the scale factor $a_{BBRW}(t, \v{r})$ is approximated in various time periods to well within an order of magnitude by

$0 \le t < t_c$

$$\mathrm{Re}\ a_{BBRW}(t, \v{r}) \cong \gamma/2 + a(t)/2 \quad (13.3.2.24)$$

$$\mathrm{Im}\ a_{BBRW}(t, \v{r}) \cong -\gamma\v{r}/2 - a(t)\v{r}/2 \quad (13.3.2.25)$$

$t_c < t$

$$\mathrm{Re}\ a_{BBRW}(t, \v{r}) \cong a(t)\{1 + (1 + \v{r}^2)\gamma^2/4\} \cong a(t) \quad (13.3.2.26)$$

$$\mathrm{Im}\ a_{BBRW}(t, \v{r}) \cong -\gamma\v{r}/2 - [\v{r}^2 + \gamma^2/(4a^2(t))]^{\frac{1}{2}}\gamma/2 \quad (13.3.2.27)$$

$t \to \infty$

$$\mathrm{Re}\ a_{BBRW}(t, \v{r}) \cong a(t)\{1 + (1 + \v{r}^2)\gamma^2/4\} \cong a(t) \quad (13.3.2.28)$$

$$\mathrm{Im}\ a_{BBRW}(t, \v{r}) \cong -\gamma\v{r} = -8.166 \times 10^{-93}\v{r} \quad (13.3.2.29)$$

13.3.3 Temperature of the Early Universe in the Generalized Robertson-Walker Metric

We will now examine the temperature of the early universe based on the results of the preceding section. For t near t = 0, we note κT depends on the blackbody scale factor of the generalized Robertson-Walker metric:[79]

$$\kappa T = \kappa T_0 / B_{BB}(t, y) \qquad (13.3.3.1)$$

and

$$B_{BB}(t, y) = 2k^{-\frac{1}{2}}M_c(1 + M_c^2 y^2)^{-1}a_{BBRW}(t, \check{r}) \qquad (13.3.3.2)$$

by eq. 13.3.2.6 using the variable

$$\check{r} \equiv M_c y \qquad (13.3.3.3)$$

for convenience. Therefore from eqs. 13.3.2.24 – 13.3.2.27 we see

$0 \le t < t_c$

$$\begin{aligned}
\kappa T_< \equiv \kappa T(0 \le t < t_c) &\cong 2\kappa T_0(1 + i\check{r})/\chi = 4\pi^{-3/2}\,M_c(1 + i\check{r}) \\
&= 2.72 \times 10^{29}(1 + i\check{r})\,\text{eV} \qquad (13.3.3.4) \\
&= 3.155 \times 10^{33}(1 + i\check{r})\,°\,\text{K}
\end{aligned}$$

$t_c < t$

$$\begin{aligned}
\kappa T_> \equiv \kappa T(t_c < t) \\
\cong \kappa T_0(a(t) + i\gamma\check{r})/a^2(t) = \kappa T_0/a(t) \qquad (13.3.3.5)
\end{aligned}$$

We note that the temperatures calculated in this section are in Extended Robertson-Walker coordinates (eq. 8.2.4.5) and not in Robertson-Walker coordinates.

Note also that the physical temperature is the absolute value of the complex temperature due to the use of the Reality group.

---

[79] Here again we remind the reader that complex values for radii, temperature and so on are transformed by the Reality group to real values – their absolute values.

13.3.4 Consistency of Y_{BB} Approximation with the Temperature near t = 0

We now address the question of whether the values found for κT are consistent with the approximations made in chapter 11 in order to obtain an expression for Y_{BB}:

$$\cos(\omega\kappa Tt) \approx 1 \qquad (8.66)$$

$$J_1(\omega\kappa Ty) \approx \omega\kappa Ty/2 \qquad (8.67)$$

The first approximation is valid if $\kappa Tt \sim 0$ since the Planck distribution factor makes the largest contribution to the integral come from small ω. The values of $\kappa T_<$ and $\kappa T_>$ show that for larger times κTt is very small:

$$\kappa T_< t_c \leq 3.42 \times 10^{-136}$$

$$\kappa T_> t \leq \kappa T_0 t/a(t) \qquad (13.3.4.1)$$

The second approximation is valid if $\kappa Ty \ll 1$ since the Planck distribution factor again makes the largest contribution to the integral come from small ω. We find the maximum values of κTy in the two time periods from eqns. 13.3.3.4 and 13.3.3.5 to be

$0 \leq t < t_c$

$$\text{MAX}(|\omega\kappa T_< y|) = |\omega\kappa T_< M_c^{-1}| = 22.27\omega \qquad (13.3.4.2)$$

which is small since the dominant part of the integration comes from small ω; and

$t_c < t$

$$\text{MAX}(|\omega\kappa T_> y|) = |\omega\kappa T_>(t_c)M_c^{-1}| = 2.35 \times 10^{-4}\omega/a(t) \qquad (13.3.4.3)$$

which is also small. Thus the Bessel function power series expansion is well approximated by its first term:

$$J_1(\omega\kappa Ty) \approx (\omega\kappa Ty/2) \qquad (13.3.4.4)$$

We conclude our approximate calculation of A(t, y) is valid for all time and for the complete range of y values: $0 \le y \le M_c^{-1}$.

13.3.5 Plots of the Scale Factor from t = 0 to the Present

We will create several plots of the scale factor from t = 0 to the present time t = 4.35 × 10^{17} s for the maximum value of y = M_c^{-1}, which corresponds to the maximum Robertson-Walker radius coordinate value r = $k^{-\frac{1}{2}}$. The approximation that we have developed for B_{BB} has been justified for times between the Big Bang and the present.

In Fig. 13.3.5.1 we show a log – log plot of the real and imaginary parts of $a_{BBRW}(t)$ vs. t using base 10 logarithms for t ∈ $[10^{-200}, 10^{20}]$ seconds. In Fig. 13.3.5.2 we plot the real and imaginary parts of a_{BBRW} vs. time from t = 0 to t = 1.2 × 10^{-246} s. In Fig. 13.3.5.3 we plot the real part of a_{BBRW}, and the Robertson-Walker scale factor a(t), vs. time in seconds in the period around 10^{-167} s. It shows the rapidity of the transition of Re a_{BBRW} from slowly rising to rapidly rising with a(t).

13.4 The Interpretation of the Complex Scale Factor

The interpretation of the complex scale factor $a_{BBRW}(t, ř)$ hinges on its role in the expression for the proper interval. The expression for the proper interval in the Robertson-Walker metric is:

$$d\tau^2 = dt^2 - R^2(t)[dr^2/(1 - kr^2) + r^2(d\theta^2 + \sin^2\theta d\varphi^2)]$$

It was extended to

$$d\tau^2 = dt^2 - A^2(t, ř)[dř^2 + ř^2(d\theta^2 + \sin^2\theta\, d\varphi^2)] \tag{10.4.9}$$

where an identification was made between the Robertson-Walker scale factor R(t) ≡ $a(t)k^{-\frac{1}{2}}$ and $a_{BBRW}(t, ř)$ through the following chain of equalities and correspondences

$$a(t) = a(t)b_0(r) = A(t, ř)/b(ř) \to A_{BB}(t, ř)(1 + ř^2)k^{\frac{1}{2}}/2 = a_{BBRW}(t, ř) \tag{13.4.1}$$

where the → relates the classical expressions on the left with the expressions on the right that embody quantum corrections due to Y-quanta blackbody radiation. Eq. 13.4.1 is based on eqns. 10.4.6, 10.4.7, 10.4.9, 10.4.11, 10.4.12, and 10.6.4a. Combining the equations behind eq. 13.4.1 we find

$$d\tau^2 = dt^2 - a_{BBRW}(t, \check{r}(r))^2[dr^2/(1 - kr^2) + r^2(d\theta^2 + \sin^2\theta d\varphi^2)] \tag{13.4.2}$$

where the relation between $\check{r}$ and r, $\check{r}(r)$, is specified by eq. 10.4.12. We call the metric in eq. 13.4.2 the Extended Robertson-Walker metric. It is equivalent to eq. 10.4.9, differing only in the definition of the radial coordinate.

Having determined the role of the complex scale factor in the metric tensor we can now physically interpret it by applying a Reality group transformation that effectively multiplies $a_{BBRW}(t, \check{r}(r))$ by a phase making it real and equal to its absolute value $|a_{BBRW}(t, \check{r}(r))|$.

We can simplify the physical interpretation without loss of generality by considering the radial coordinate of eq. 10.4.9 to be

$$r_{GRW} = A_{BB}\check{r} \tag{13.4.3}$$

$$= r_r + ir_i \tag{13.4.4}$$

where

$$A_{BB}(t, y) = 2a_{BBRW}(t, \check{r})/[(1 + M_c^2 y^2)k^{\frac{1}{2}}] \tag{8.5.2.4}$$

Thus

$$r_{GRW} = 2a_{BBRW}(t, \check{r})\check{r}/[(1 + \check{r}^2)k^{\frac{1}{2}}] \tag{13.4.5}$$

Applying a Reality group transformation we obtain the physically measurable, real-valued radius

$$r_{GRWphysical} = 2|a_{BBRW}(t, \check{r})\check{r}/[(1 + \check{r}^2)k^{\frac{1}{2}}]| \tag{13.4.5a}$$

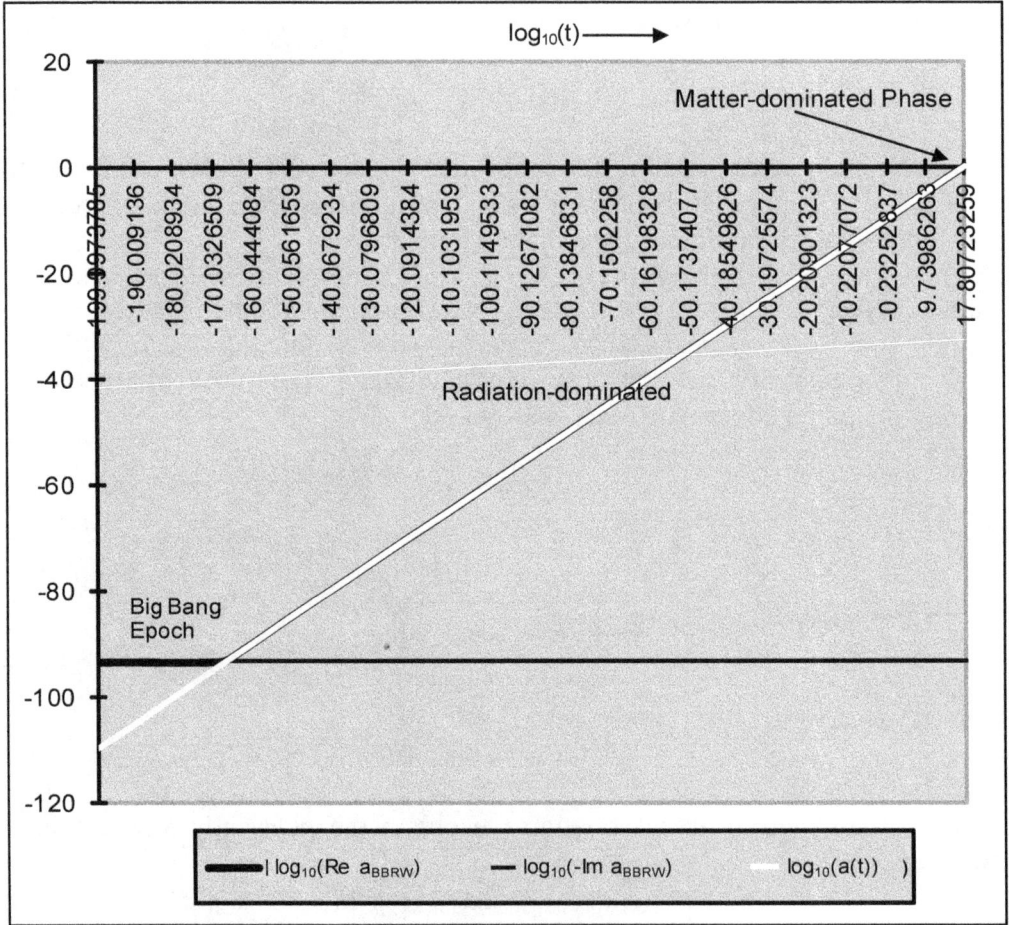

Figure 13.3.5.1. A log-log (base 10) plot of the real and imaginary parts of a_{BBRW} and the Robertson-Walker scale factor a(t) versus the log (base 10) of time in seconds. Note the imaginary part of a_{BBRW} is very slowly growing. The real part of a_{BBRW} is growing slowly until $t_c \approx 10^{-165}$ s and thereafter equals a(t) to good approximation.

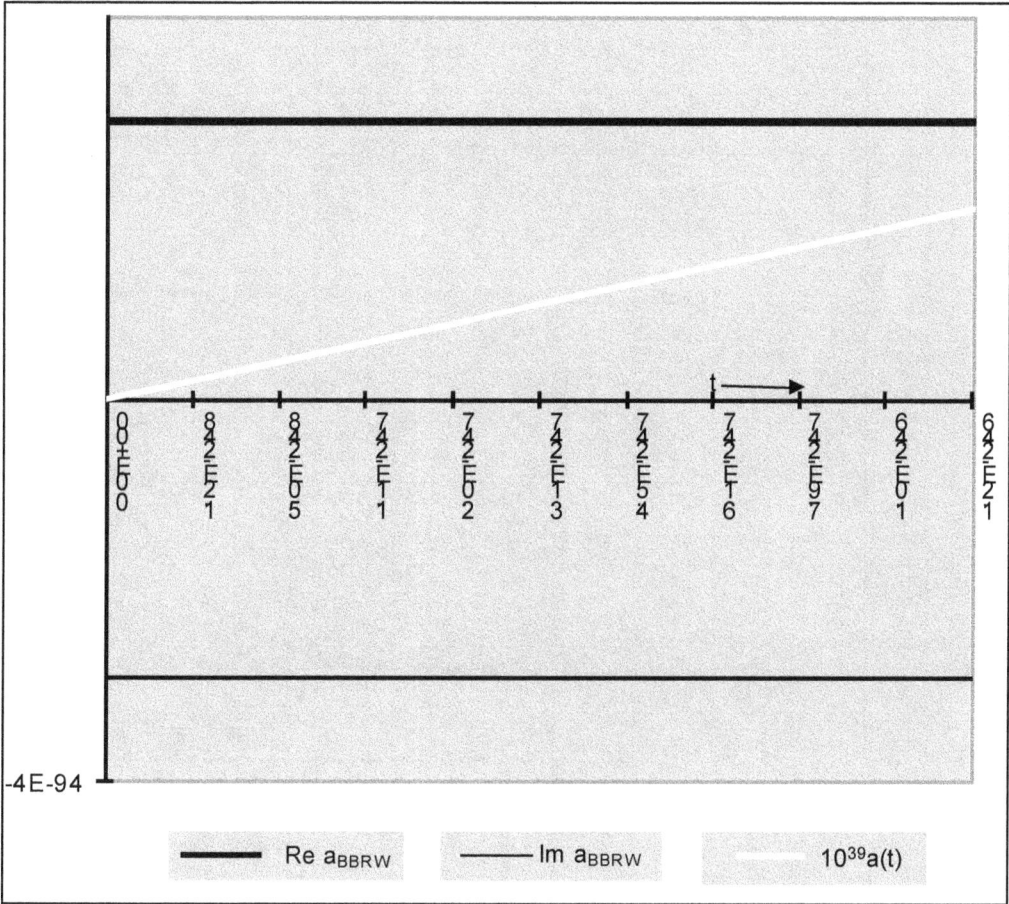

Figure 13.3.5.2. A plot of the real and imaginary parts of a_{BBRW}, and $10^{39} \times a(t)$, versus time from $t = 0$ to $t = 1.2 \times 10^{-246}$ s. Note they are slowly varying and well behaved in the neighborhood of $t = 0$ with only $a(t)$ having the value of zero. "E" indicates a power of ten (for example: $2.0E\text{-}248 = 2.0 \times 10^{-248}$ s).

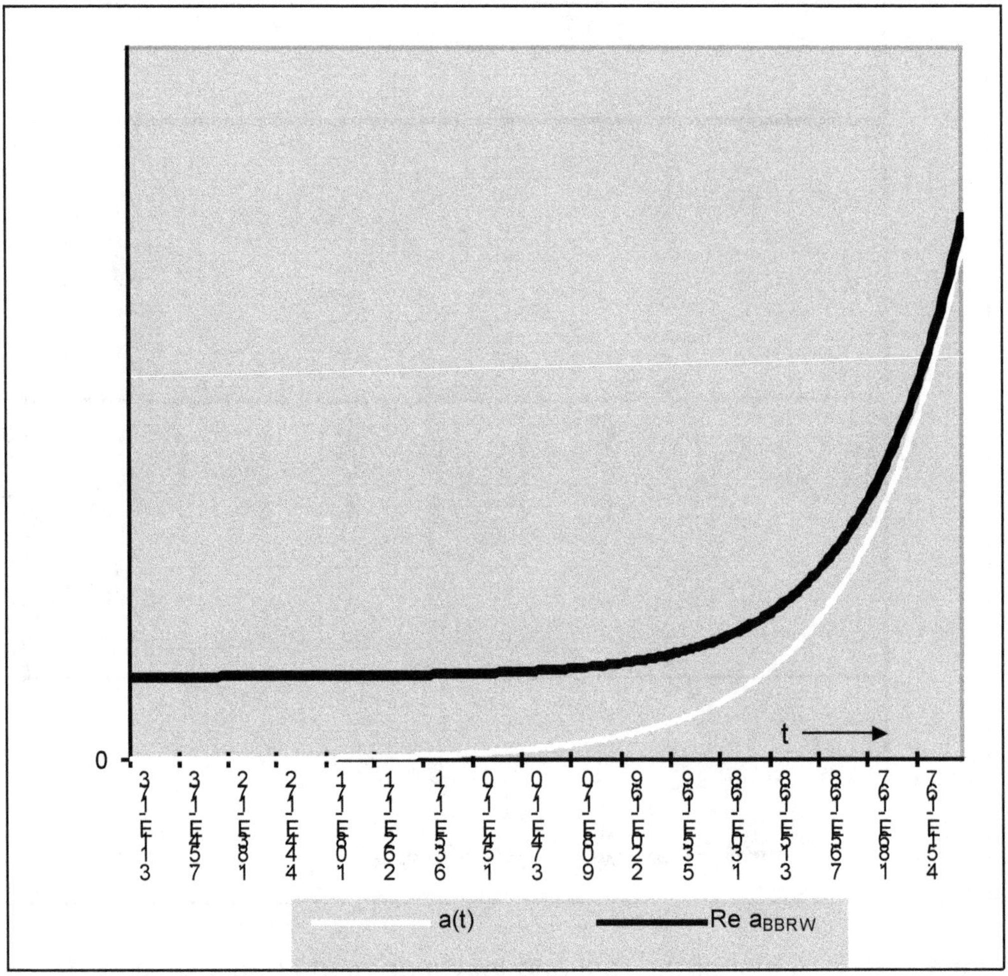

Figure 13.3.5.3. A plot of the real part of a_{BBRW} and a(t) vs. time in seconds before the time 10^{-165} s. Note Re a_{BBRW} quickly changes from slowly growing to growing like a(t).

13.5 Time Evolution of the Hubble Rate

The Hubble rate is one of the linchpins of modern cosmology. It is determined by Einstein's equation eq. 13.2.1.1 when written in the form:

$$H(t) = \dot{a}/a = [H_0^2(\Omega_\gamma/a^4(t) + \Omega_m/a^3(t) + \Omega_\Lambda) - k/a^2(t)]^{1/2} \qquad (13.5.1)$$

At small times, in the radiation-dominated phase, eq. 13.5.1 can be approximated by

$$H(t) \cong H_0\Omega_\gamma^{1/2}/a^2(t) \qquad (13.5.2)$$

If we define a Hubble rate $H(t)$ using $a_{BBRW}(t, \check{r})$ then

$$H_{BBRW}(t, \check{r}) = |\dot{a}_{BBRW}(t, \check{r})/a_{BBRW}(t, \check{r})| \equiv |\dot{A}_{BBRW}(t, \check{r})/A_{BBRW}(t, \check{r})|$$
$$= |[H_0^2(\Omega_\gamma/a_{BBRW}^4(t, \check{r}) + \Omega_m/a_{BBRW}^3(t, \check{r}) + \Omega_\Lambda) - k/a_{BBRW}^2(t, \check{r})]^{1/2}| \qquad (13.4.3)$$

$H_{BBRW}(t, \check{r})$ is the same as $H(t)$ until we reach the first instants of the universe, which we have called the Big Bang Epoch., Then we find

$0 \le t < t_c$

$$H_{BBRW}(t, \check{r}) \cong H_0\Omega_\gamma^{1/2}/|a_{BBRW}(t, \check{r})|^2 \qquad (13.4.4)$$

where

$$\text{Re } a_{BBRW}(t, \check{r}) \cong \gamma/2 + a(t)/2 \qquad (13.3.2.24)$$

and

$$\text{Im } a_{BBRW}(t, \check{r}) \cong -\gamma\check{r}/2 - a(t)\check{r}/2 \qquad (13.3.2.25)$$

Substituting in eq. 13.4.4 we find

$$H_{BBRW}(t, \check{r}) \cong 4H_0\Omega_\gamma^{1/2}|[1 - \check{r}^2 + 2i\check{r}]/[(\gamma + a(t))^2(1 + \check{r}^2)^2]| \qquad (13.4.5)$$

$|H_{BBRW}(t, \check{r})|$ is a real physical number in this range. Thus space has a Hubble rate that is both space and time dependent in the Big Bang Epoch.

13.6 Hubble Constant in the Big Bang Metastate at t = 0 and t = t_c

13.6.1 At Time t = 0

At t = 0 we find $H_{BBRW}(0, \check{r})$ is finite unlike the radiation-dominated Hubble rate (eq. 13.5.2):

$$H_{BBRW}(0, \check{r}) \cong 4H_0\Omega_{,\gamma}^{\frac{1}{2}}|[1 - \check{r}^2 + 2i\check{r}]/[\gamma^2(1 + \check{r}^2)^2]| \qquad (13.4.6)$$

At the "edge" of the universe the Hubble rate is

$$H_{BBRW}(0, 1) \cong 2H_0\Omega_{,\gamma}^{\frac{1}{2}}/\gamma^2 = 5.8 \times 10^{198} \text{ s}^{-1} \qquad (13.4.7)$$
$$= 1.79 \times 10^{218} \text{ km s}^{-1} \text{ Mpc}^{-1}$$

(compared to the current Hubble constant of 100h km s^{-1} Mpc^{-1}), and is solely due to the imaginary part of $4H_0\Omega_{,\gamma}^{\frac{1}{2}}[1 - \check{r}^2 + 2i\check{r}]/[(\gamma + a(t))^2(1 + \check{r}^2)^2]$ since the real part is zero. At $\check{r} = 0$,

$$H_{BBRW}(0, 0) \cong 4H_0\Omega_{,\gamma}^{\frac{1}{2}}1/\gamma^2 = 2H_{BBRW}(0, 1) \qquad (13.4.8)$$

is more rapidly expanding.

Thus we consistently avoid the divergences that appear at t = 0 in the Standard Cosmological Model. Its radiation-dominated phase's Hubble rate is $(2t)^{-1}$ which diverges at t = 0.

13.6.2 At Time t = t_c

At the "edge" of the universe the enormous Hubble rate at the exit time $t_c = 1.26 \times 10^{-165}$ s from the Big Bang metastate is[80]

$$H_{BBRW}(t_c, 1) \cong 4H_0\Omega_{,\gamma}^{\frac{1}{2}}|i|/[2(\gamma + a(t_c))^2]| \qquad (13.4.9)$$
$$= 2 H_0\Omega_{,\gamma}^{\frac{1}{2}}/(\gamma + \gamma^2) \cong 2 H_0\Omega_{,\gamma}^{\frac{1}{2}}/\gamma = 3.7 \times 10^{106} \text{ s}^{-1}$$
$$= 1.14 \times 10^{126} \text{ km s}^{-1} \text{ Mpc}^{-1}$$

---

[80] Eq. 13.3.2.15.

compared to the current Hubble constant of $100h$ km s^{-1} Mpc^{-1}.
At $\check{r} = 0$, the "center" of the Big Bang has

$$H_{BBRW}(t_c, 0) \cong H_0\Omega_\gamma^{1/2}/\gamma^2 = 2.9 \times 10^{198} \text{ s}^{-1} \qquad (13.4.10)$$
$$= 8.95 \times 10^{217} \text{ km s}^{-1} \text{ Mpc}^{-1}$$

using eq. 13.3.2.14. The "center" *is more rapidly expanding by a factor of the order of 10^{93}, which is suggestive of a developing much less dense region.*

13.6.3 Hubble Constant Decrease in the Metastable Big Bang State

In the metastable state lasting 10^{-165} s, at the "edge" of the expanding universe, the Hubble constant has shrunk enormously from 5.8×10^{198} s^{-1} to 3.7×10^{106} s^{-1} — a decrease of a factor of 10^{92} — surfacing again Eddington's 90 orders of magnitude conjecture. The origin of the 90 orders of magnitude appears to be in a constant that combines standard constants appearing in our Two-Tier Quantum Gravity based on black body, the Big Bang expansion and the current CMB temperature. The Beginning is thereby directly related to the Present:

$$\gamma = x/\varpi = 6.38 \times 10^{-93} = \pi^{3/2} \text{ k}^{1/2}\kappa T_0/(2M_c^2) \qquad (13.4.11)$$

Eq. 13.4.11 uses eqs. 13.3.1.2 and 13.3.2.2, and the CMB temperature in.eq. 13.3.1.1.

13.7 Creation of a Void in the Metastable State due to a Radially Varying Hubble Constant

Sections 13.6.1 and 13.6.2 show that at $t = 0$ the Hubble "constant" is a factor of 2 greater at $\check{r} = 0$ than at the edge of the universe ($\check{r} = 1$), and at $t = t_c$ (the end of the metastable state life) the Hubble "constant" is a factor of the order of 10^{93} greater at $\check{r} = 0$ than at the edge of the universe ($\check{r} = 1$). The vast disparity in Hubble "constant" values at the center compared to the edge of the universe suggests the center will rapidly go to lower density creating a *void. Thus one can expect a "Hubble bubble" generated during the metastable state.*

Today we see numerous voids in the universe.[81] The largest is the spherical KBC supervoid (containing the Milky Way) with a 2 billion light year diameter. Another supervoid is the "Giant void" (Canes Venatici) with a diameter of 1.3 light years. There is also the WMAP "cold spot" void with a diameter of 120 Megaparsecs.

The Hubble Constant within this void is larger due to the attraction of mass-energy external to the void just as we see above when comparing the center to the edge of the Metastable state.

The metastate "void" appears to be a precursor to the later voids seen now. Voids may also be the result of the fluctuation in universe size around the time of the Big Dip. See Section 4 for more details.

13.8 Energy Density in the Big Bang Metastable State

An estimate of the density of energy in the Big Bang metastable state can be obtained as follows. At small times, in the radiation-dominated phase, eq. 13.5.1 can be approximated by

$$a'/a = H(t) \cong H_0\Omega_\gamma^{1/2}/a^2(t) \cong H_{BBRW}(t, \check{r}) = H_0\Omega_\gamma^{1/2}/a_{BBRW}^2(t) \quad (13.5.2)$$

from eq. 13.4.5. The "Big Bang" regime Einstein Equation is

$$a'_{BBRW}^2 - 8\pi G\rho_{BBRW}a_{BBRW}^2/3 = -k \quad (13.7.1)$$

implying

$$\rho_{BBRW} = (3/8\pi G)[(a'/a) + k]$$
$$= (3/8\pi G)[H_{BBRW}(t, \check{r}) + k] \quad (13.7.2)$$

Since

$$H_{BBRW}(t, \check{r}) \cong H_0\Omega_\gamma^{1/2}/|a_{BBRW}(t, \check{r})|^2 \quad (13.4.4)$$

and

$$Re\ a_{BBRW}(t, \check{r}) \cong \gamma/2 + a(t)/2 \quad (13.3.2.24)$$
$$Im\ a_{BBRW}(t, \check{r}) \cong -\gamma\check{r}/2 - a(t)\check{r}/2 \quad (13.3.2.25)$$

we find

[81] Zehavi *et al*, Astrophysical Journal **503**, 483 (1998); There are differing results on the Hubble Bubble question such as in Moss *et al*, Phys. Rev. **D83**, 103515 (2011) and references therein.

$$\rho_{BBRW} \, (t = 0, \check{r} = 0) \cong (3/8\pi G)[4H_0\Omega_\gamma^{\,\frac{1}{2}}/\gamma^2 + k] \cong (3/2\pi\gamma^2 G)H_0\Omega_\gamma^{\,\frac{1}{2}} \qquad (13.7.3)$$
$$\cong 2.9 \times 10^{194} \ \text{gm/cm}^3 = 1.63 \times 10^{218} \ \text{GeV/cm}^3$$

$$\rho_{BBRW} \, (t = 0, \check{r} = 1) \cong (3/8\pi G)[2H_0\Omega_\gamma^{\,\frac{1}{2}}/\gamma^2 + k] \cong (3/4\pi\gamma^2 G)H_0\Omega_\gamma^{\,\frac{1}{2}} \qquad (13.7.4)$$
$$\cong 1.45 \times 10^{194} \ \text{gm/cm}^3 = 8.16 \times 10^{217} \ \text{GeV/cm}^3$$

At t = t_c:

$$\rho_{BBRW} \, (t = t_c, \check{r} = 0) \cong 3/(8\pi G)[H_0\Omega_\gamma^{\,\frac{1}{2}}/\gamma^2 + k] \cong 3/(8\pi\gamma^2 G)H_0\Omega_\gamma^{\,\frac{1}{2}} \qquad (13.7.5)$$
$$\cong 3.7.26 \times 10^{193} \ \text{gm/cm}^3 = 4.08 \times 10^{217} \ \text{GeV/cm}^3$$

$$\rho_{BBRW} \, (t = t_c, \check{r} = 1) \cong 3/(8\pi G)[H_0\Omega_\gamma^{\,\frac{1}{2}}/(2\gamma^2) + k] \cong 3/(16\pi\gamma^2 G)H_0\Omega_\gamma^{\,\frac{1}{2}} \qquad (13.7.6)$$
$$\cong 3.63 \times 10^{193} \ \text{gm/cm}^3 = 2.04 \times 10^{217} \ \text{GeV/cm}^3$$

14. From t = t_T to the Present: Scale Factor and Hubble Constant

Many years ago Hoyle and Narlikar, and others,[82] proposed a steady state theory of the evolution of the universe in which mass-energy appeared uniformly throughout the universe. They had no convincing explanation for the origin of the influx of mass-energy. In this book we provide a detailed explanation for the source of the influx: the Megaverse. The surface tension force (gravitational) of our universe essentially absorbs mass-energy from the surrounding Megaverse.

In this chapter we will consider the time period from t = t_T =380,000 years to the present during which the influx occurs.[83]

14.1 Influx Epoch Mass-Energy Density and Scale Factor Calculation

Our first concern is to determine $\rho_{tot}(t)$ during the influx time period. We must calculate the total mass-energy density as a function of time including the accumulated total mass-energy influx from t = t_T.

Earlier we found that the derivative of $\rho_{tot}(t)$ (eq. 9.6) is

$$d\rho_{tot}(t)/dt = \rho_{crit}d\Omega_H(t)/dt + C_M k^{\frac{1}{2}}\rho_{tot}(t)/a(t) \tag{9.9}$$

using $R = k^{-\frac{1}{2}}a(t)$ and

$$C_M = C_T G\rho_{Mega} \tag{9.8}$$

$$\Omega_H(t) = \Omega_\gamma(t) + \Omega_\Lambda(t) + \Omega_m(t) + \Omega_d(t) \tag{9.10}$$

---

[82] H. Bondi and T. Gold, M.N.R.A.S. **108**, 252 (1948); F. Hoyle, M.N.R.A.S. **108**, 372 (1948); F. Hoyle and J. V. Narlikar, Proc. R. Soc. London **A290**, 162 (1966) and references therein.
[83] We selected this time for easy comparison to the known Hubble Constant data.

The solution of eq. 9.9 is

$$\rho_{tot}(t) = \rho_{crit} e^{F(t)} \int_{t_T}^{t} dt'' \, e^{-F(t'')} \, d\Omega_H(t'')/dt'' \qquad (14.1)$$

where

$$F(t) = C_M k^{\frac{1}{2}} \int_{t_T}^{t} dt'/a(t') \qquad (14.2)$$

The above equations for ρ_{to} together with the Einstein equation;

$$\dot{a}^2 - 8\pi G \rho_{tot} a^2/3 = -k \qquad (13.1.2)$$

can be solved simultaneously for the Influx mass-energy density $\Omega_T(t)$, the total density $\rho_{tot}(t)$ as a function of time, and the scale factor a(t) given the values of c_M, T_c, and $\Omega_H(t)$.

14.2 Decomposition of the Mass-Energy Density ρ_{tot} Equation

Eq. 9.9 can be separated into four equations, one equation for each of $\Omega_\gamma(t)$, $\Omega_\Lambda(t)$, $\Omega_m(t)$, $\Omega_d(t)$ and $\Omega_T(t)$. Letting $\Omega_i(t)$ represent each of these densities we find the identities

$$d\Omega_i(t)/dt = d\Omega_i(t)/dt \qquad (14.3)$$

for the first four energy densities of the separations using $\rho_i(t) = \rho_{crit}\Omega_i(t)$, and

$$d\Omega_T(t)/dt = C_M k^{\frac{1}{2}}[\Omega_H(t) + \Omega_T(t)] \, /a(t) \qquad (14.4)$$

for $\Omega_T(t)$. Eq.14.3 indicates that the energy influx does not change the dynamical equations for $\Omega_H(t)$ but does determine $\Omega_T(t)$.

The solution to eq. 14.4 is

$$\Omega_T(t) = C_M k^{\frac{1}{2}} e^{F(t)} \int_{t_T}^{t} dt'' \, e^{-F(t'')} \Omega_H(t'')/a(t'') \qquad (14.5)$$

14.3 Megaverse Mass-Energy Densities

The constant Megaverse mass-energy densities will be assumed to be in the same proportions as in our universe at the present time $t = t_{now}$ with

$$\rho_{Mega} = \rho_{Mega0}(\Omega_{\gamma M} + \Omega_{\Lambda M} + \Omega_{mM} + \Omega_{dM})$$

where ρ_{Mega0} is a Megaverse constant "critical density" and[84] the Megaverse energy densities (which are not used in further computations in this section) are assumed to be

$$\Omega_{\gamma M} = 5.38 \times 10^{-5} \qquad (14.6)$$
$$\Omega_{\Lambda M} = 0.692$$
$$\Omega_{mM} = 0.046$$
$$\Omega_{dM} = 0.268$$

as currently in our universe.

---

[84] M. Tanabashi *et al op. cit.*

15. The Big Bang Epoch

No great thing is created suddenly.
<u>*Discourses*</u> *- Epictetus*

15.1 The $t = 0$ Big Bang Scenario

The Two-Tier cosmological theory that we have developed in preceding chapters differs dramatically from the Standard Cosmological Model in the Big Bang Epoch and yet smoothly melds into the Standard Cosmological Model in the Expanding Universe and Exploding Universe epochs in chapter 13. Chapter 14 revises the picture significantly when an influx of Megaverse matter is introduced.

The universe has a finite size, temperature and density at the point of the Big Bang which we define to be the time t = 0. The universe grows slowly for a period of time (until roughly 10^{-165} s) that we call the Big Bang metastable state. In this time period we see a very hot, very dense, macroscopic conglomeration of Y quanta, radiation and elementary particles that coexist with each other with non-singular interactions. These particles are not localized. Each particle can be said to occupy the entire universe since the size of the universe is infinitesimal compared to any particle's Compton radius (if it had one). The universe has an almost classical Robertson-Walker type of metric. In particular, it has an Extended Robertson-Walker metric with quantum effects due to a classical Y-quanta blackbody radiation field (the inflatons?) that is both the source of the metastability of the universe at t = 0 and the source of infinitesimal imaginary spatial dimensions that are comparable in extent with the size of the real spatial dimensions of the universe during the Big Bang Epoch.

As the universe expands due to the Y quanta energy it appears that enormous amounts of gravitational energy are also released since the Two-Tier gravitational potential is zero at r = 0 and has a minimum around $r \approx 10^{-33}$ cm. Thus we view the universe in the Big Bang Epoch as a "slowly" expanding metastable state. (This state is comparable to the metastable false vacuum state in inflation theories – but no scalar bosons are needed – the Y quanta play that role. The combination of gravitation and an effectively classical Y field serve to generate the metastable initial state of the Big Bang.

We can summarize some of our model's features (many of which are calculated later in this chapter) with:

1. The universe is a macroscopic object in terms of content and as such can be described by classical physics – modified by quantum effects.

2. Quantum fluctuations do not play any significant role in the Big Bang Epoch because of the nature of Two-Tier quantum field theory. For example, the quantum fluctuations of the quantized gravitational field were shown to be zero in chapter 7 of Blaha (2004):

$$<0|h_{\mu\nu}(X)h_{\alpha\beta}(X)|0> = \int d^3p\ b'_{\mu\nu\alpha\beta}(p)\ e^{-p^ip^j{}_\Delta T_{ij}(0)}/[(2\pi)^3 2\omega_p] = 0 \quad (7.3.8.3.3)$$

3. All Two-Tier quantum fields also have zero quantum fluctuations. This behavior holds whether they are quantized in flat space or in a curved space. Thus quantum fluctuations (or foam) are not an issue in Two-Tier theories.

4. The radiation and matter in the Big Bang Epoch produces, in effect, a classical Y-quanta blackbody spectrum that modifies the nature of space making it complex with comparable real and imaginary spatial parts. This effect appears in the spatial scale factor of the Extended Robertson-Walker metric described in previous chapters. A Reality group transformation transforms spatial coordinates to real physical values.

5. The usual scale factor a(t) is determined by the conventional Standard Cosmological Model classical Einstein equation with an appropriate choice of energy density and pressure, since its source is the macroscopic energy density.

6. The radius of the universe (with both real and imaginary parts) at the beginning of the Big Bang Epoch is

$$r_{universe}(0) = a_{BBRW}(0)/k^{1/2} = \gamma(1 - i)/(2k^{1/2}) = 4.3(1 - i) \times 10^{-65} \text{ cm} \quad (15.1.1)$$

The physical radius is the absolute value of $r_{universe}(0)$. The confinement of particles to a radius of this size means that they cannot be considered to be localized but, rather, they are spread over the entire volume of the universe.

7. The small size of the universe implies Two-Tier potentials between particles, and particle propagators, are effectively zero for all particles. Consequently the universe is in a metastable state. Some idea of the relative potential energy of Two-Tier QFT particles vs. ordinary QFT particles can be gleaned by comparing a standard Newton-Coulomb type of potential (knowing that it would be modified in strong gravitational fields if they were present)

$$V_{std} = 1/r \quad (15.1.2)$$

with the Two-Tier potential at short distances (below the Planck scale):

$$V_{tt} = 2\sqrt{\pi} \, M_c^2 r \quad (15.1.3)$$

obtained from eq. 7.3.9.3 of Blaha (2004). We have not displayed the coupling constant in eqs. 15.1.2 and 15.1.3. At $r = r_{universe}(0) = 4.3 \times 10^{-65}$ cm

$$V_{std} = 4.57 \times 10^{59} \text{ eV} \quad (15.1.4)$$

$$V_{tt} = 2\sqrt{\pi} \, M_c^2 r = 0.00115 \text{ eV} \quad (15.1.5)$$

Thus the Two-Tier potential between particles is negligible at distances up to the radius of the Big Bang Epoch universe at t = 0 since the Heisenberg Uncertainty Principle, when applied using the "diameter" of the universe as the uncertainty in position, implies the uncertainty in a particle's energy, and thus the scale of particle energies, is (coincidentally) of the order of 10^{59} eV. (See the discussion in the following sections.)

8. In view of the above points it is reasonable, and self-consistent, to use the Extended Robertson-Walker metric that we have developed in the preceding chapters.

9. The t = 0 universe is very dense with an energy density of the order of 10^{339} g/cm^3 of particles with negligibly small, non-singular (as particle separation goes to zero) interactions. (See the discussion below.)

10. The Big Bang Epoch is a metastable state. Due to the form of the Two-Tier gravitational potential it has a much higher gravitational potential energy at t = 0 than when the radius of the universe is about 10^{-33} cm. **In fact, Gravity is a repulsive force (anti-gravity!) at distances less than 9.08×10^{-34} cm.** (Fig. 7.3.9.3 of Blaha (2004) is reproduced on the next page for the reader's convenience.)

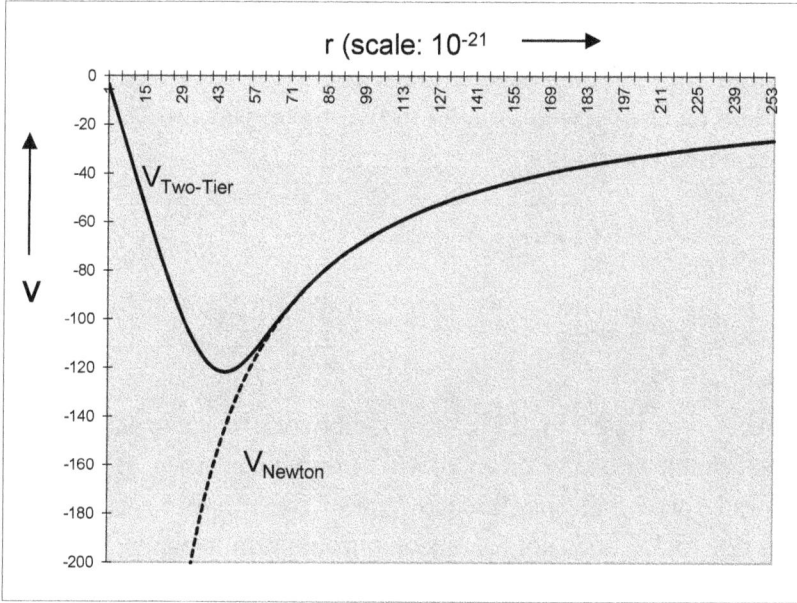

Figure 15.1.1. A plot of the Two-Tier gravitational potential (solid line) and the Newtonian gravitational potential (dashed line). Note anti-gravity at short distances. See Fig. 7.3.9.3 of Blaha (2004) for more details.

The Two-Tier gravitational potential between two particles has a minimum at (eq. 7.3.9.5 of Blaha (2004))

$$r_{MIN} = \pi^{-\frac{1}{2}} M_c^{-1} = 9.08 \times 10^{-34} \text{ cm} \tag{15.1.6}$$

At the minimum the gravitational $V_{\text{Two-Tier}}$ has the value:

$$V_{\text{Two-TierMIN}} = -.8427 \sqrt{\pi} \, GM_c = -1.22 \times 10^{-28} \text{ eV}^{-1} \tag{15.1.7}$$

Since the energy of the universe is confined within a radius of 10^{-65} cm at t = 0 there is a tremendous release of gravitational potential energy as the universe expands to 10^{-33} cm and beyond. This energy is converted initially into kinetic energy and can be viewed as helping fuel the expansion. A crude approximation to this released energy is

$$\Delta E \sim (\rho V_{universe})^2 V_{Two\text{-}TierMIN}$$

$$\cong (1.6 \times 10^{212} \text{ g/cm}^3 \times 1.8 \times 10^{-100} \text{ cm}^3)^2 \times 1.22 \times 10^{-28} \text{ eV}^{-1}$$
$$\cong 3 \times 10^{262} \text{ eV} \tag{15.1.8}$$

The energy estimate (eq. 15.1.8) is far beyond the current total energy of the universe which is of the order of 10^{57} eV. Since energy is not conserved (It is considered somewhat undefined by many General Relativists.) as the universe evolves eq. 15.1.8 does not represent a problem in itself.

11. The temperature of the real part of the universe at t = 0 is

$$\kappa T \approx 2 \times 10^{89} \text{ eV} \tag{15.1.9}$$

as shown later.

15.2 The Radius of the Universe in the Big Bang Epoch

In this section we will consider the radius of the universe at the point of the Big Bang (t = 0) and its relatively slow growth during the Big Bang Epoch (t ≤ $t_c \cong 10^{-165}$ s).

The current radius[85] of the universe is 4.314×10^{28} cm. Motivated by that finding we assume the current radius of the universe in the Robertson-Walker model is:

Assumption: $r_{universe}(t_{now}) = a(t_{now})/k^{1/2} \approx 4.314 \times 10^{28}$ cm   (15.2.1)

---

[85] M. Tanabashi *et al* (Particle Data Group) *op. cit.*

It is reasonable to define the radius of the universe at earlier times correspondingly. In our Two-Tier blackbody model for the universe it is

$$r_{universe}(t) = a_{BBRW}(t, \check{r} = 1)/k^{\frac{1}{2}} \qquad (15.2.2)$$

with

$$a_{BBRW}(t, \check{r}) = \{-i\gamma\check{r} + a(t) + [\gamma^2 - 2i\gamma\check{r}a(t) + a^2(t)]^{\frac{1}{2}}\}/2 \qquad (15.3.2.19)$$

Thus the radius of the universe is complex and given by

$$r_{universeBBRW}(t, \check{r}=1) = \{-i\gamma + a(t) + [\gamma^2 - 2i\gamma a(t) + a^2(t)]^{\frac{1}{2}}\}/(2k^{\frac{1}{2}}) \qquad (15.2.3)$$

The physical radius of the universe is the absolute value of eq. 15.2.3. It is calculated by applying a Reality group transformation.

The real and imaginary parts of $r_{universeBBRW}(t)$ are:

$$\text{Re } r_{universeBBRW}(t) = \{a(t) + [R(1 + \cos \psi)/2]^{\frac{1}{2}}\}/(2k^{\frac{1}{2}}) \qquad (15.2.4a)$$

and

$$\text{Im } r_{universeBBRW}(t) = \{-\gamma - [R(1 - \cos \psi)/2]^{\frac{1}{2}}\}/(2k^{\frac{1}{2}}) \qquad (15.2.4b)$$

where

$$R = [(\gamma^2 + a^2(t))^2 + 4\gamma^2 a^2(t)]^{\frac{1}{2}} \qquad (15.3.2.12)$$

and

$$\cos \psi = (\gamma^2 + a^2(t))/R \qquad (15.3.2.13)$$

From the behavior of $a_{BBRW}(t)$ displayed in Figs. 15.3.5.1 – 15.3.5.3 we see that it is well approximated by eqns. 15.3.2.24 – 15.3.2.29 with $\check{r} = 1$. Therefore for $t < t_c$

$t < t_c$

$$\text{Re } r_{universeBBRW}(0 \leq t < t_c) \cong (\gamma + a(t))/(2k^{\frac{1}{2}}) \qquad (15.2.5a)$$

$$\text{Re } r_{universeBBRW}(0) \cong \gamma/(2k^{\frac{1}{2}}) = 4.278 \times 10^{-65} \text{ cm} \qquad (15.2.5b)$$

$$\text{Im } r_{\text{universeBBRW}}(0 \leq t < t_c) \cong -(\gamma + a(t))/(2k^{\frac{1}{2}}) \qquad (15.2.6a)$$

$$\text{Im } r_{\text{universeBBRW}}(0) \cong -\gamma/(2k^{\frac{1}{2}}) = -4.278 \times 10^{-65} \text{ cm} \quad (15.2.6b)$$

from eqns. 15.3.2.24 and 15.3.2.25. Since the Planck length is 1.61×10^{-33} cm we see the real part of the radius of the universe at $t = 0$ (until $t \approx t_c$) is over thirty orders of magnitude smaller than the Planck length.

If we use eqns. 15.3.2.28 and 15.3.2.29 then the radius at present is

$t = t_{now}$

$$r_{\text{universeBBRW}}(t_{now}) = a(t_{now})/k^{\frac{1}{2}} - i\gamma/k^{\frac{1}{2}} \qquad (15.2.7)$$

with

$$\text{Re } r_{\text{universeBBRW}}(t_{now}) = a(t_{now})/k^{\frac{1}{2}} = 1.34 \times 10^{28} \text{ cm} \qquad (15.2.8)$$

as above (eq. 15.2.1), and

$$\text{Im } r_{\text{universeBBRW}}(t_{now}) = -\gamma/k^{\frac{1}{2}} = -8.55 \times 10^{-65} \text{ cm} \qquad (15.2.9)$$

The ratio of the current, and the $t = 0$ real parts of the, radius of the universe is huge:

$$r_{\text{universe}}(t_{now})/(\text{Re } r_{\text{universeBBRW}}(0)) = 1.008 \times 10^{93} \qquad (15.2.10)$$

The real part of the universe has expanded dramatically while the imaginary part has remained almost constant in size. The physical radius of the universe for $t < t_c$ and $t = t_{now}$ are the absolute values of the complex radius values above.

The value of the Robertson-Walker radius coordinate at points *within* the universe is specified by

$$r_{RW}(t, \check{r}) = a_{BBRW}(t, \check{r})r \qquad (15.2.11)$$

The coordinate r is related to the $\check{r}$ coordinate by

$$r = 2k^{-\frac{1}{2}}\check{r}(1 + \check{r}^2)^{-1} \qquad (10.4.11)$$

Thus

$$r_{RW}(t, \check{r}) = 2k^{-\frac{1}{2}}\check{r}a_{BBRW}(t, \check{r})(1 + \check{r}^2)^{-1} \qquad (15.2.12)$$

The real and imaginary parts of $a_{BBRW}(t, \check{r})$ are specified in eqns. 15.3.2.24 – 15.3.2.29. Again we note the absolute values of eqns. 15.2.11-15.2.12 are the physical values of the Extended Robertson-Walker radius coordinate.

15.2.1 Localization of Particles at t = 0

The radius of the universe in the neighborhood of $t = 0$ is approximately 4.278×10^{-65} cm. The particles within that incredibly small universe are still the "wave-particles" that we are familiar with within the framework of quantum mechanics. As such, the position and momentum of the particles must satisfy the Heisenberg Uncertainty Condition:

$$\Delta p \Delta x \geq \hbar \qquad (15.2.1.1)$$

where $\hbar$ is Planck's constant divided by 2π. In view of the extraordinary small size of the universe – much smaller than the Compton wavelength of any known massive particle – the "spread" (uncertainty) in a particle's position is set by the radius of the universe:

$$\Delta x \approx 2 \, |r_{universeBBRW}| \qquad (15.2.1.2)$$

Thus the "spread" in the momentum of a particle (the "size" of the region in momentum space where the Fourier transform of the particle's wave function is large) is

$$\Delta p \approx \hbar/\Delta x \approx \hbar/[2|\text{Re } r_{universeBBRW}|] = 1.6 \times 10^{59} \text{ eV} = 1.3 \times 10^{31} M_{Planck} \qquad (15.2.1.3)$$

One can only view the particles in the universe in the neighborhood of $t = 0$ as spread across the entire universe. They are entirely <u>unlocalized</u> within the universe from a quantum viewpoint. Since all forces are non-singular in the very small universe at the beginning of time we can view the particles as co-resident in the same spatial region that constitutes the universe. (Simply put, they interpenetrate each other.) Thus the question of particle horizons and the homogeneity of the universe in the Beginning are irrelevant. *The universe today does not scale down to a micro-universe of the same sort as our universe in the neighborhood of t = 0.*

15.3 Volume of Big Bang Metastate

Using the Extended Robertson-Walker model we find the total volume

$$V(t) = 2\pi^2 R^3(t) k^{-\frac{3}{2}}$$
$$= 2\pi^2 \, r_{universeBBRW}^3 (t, \check{r}=1) \tag{15.3.1}$$

Earlier we found

$$r_{universeBBRW}(t, \check{r}=1) = \{-i\gamma + a(t) + [\gamma^2 - 2i\gamma a(t) + a^2(t)]^{\frac{1}{2}}\}/(2k^{\frac{1}{2}}) \tag{15.2.3}$$

Therefore

$$V(0) = \pi^2 \gamma^3 |(1-i)|^3/(4k^{3/2}) \tag{15.3.2}$$
$$= 4.37 \times 10^{-192} \text{ cm}^3$$

and

$$V(t_c) = \pi^2 \gamma^3 |1 - i + \sqrt{2}(1-i)^{\frac{1}{2}}|^3/(4k^{3/2}) \tag{15.3.3}$$
$$= \pi^2 \gamma^3 |1 - i + \sqrt{2}(1-i)^{\frac{1}{2}}|^3/(4k^{3/2})$$
$$= 2.6398 \times 10^{-191} \text{ cm}^3$$

using eq. 13.3.2.14.

15.4 Big Bang Epoch Total Y-Field Expansion Energy of Universe

We can estimate the total Y-field expansion energy[86] of the universe at t= 0 and t = t_c using

$$\rho_{BBRW}(t = 0, \check{r} = 0) \cong (3/8\pi G)[4H_0\Omega_\gamma^{\frac{1}{2}}/\gamma^2 + k] \cong (3/2\pi\gamma^2 G)H_0\Omega_\gamma^{\frac{1}{2}} \tag{13.7.3}$$

---

[86] In the Big Bang metastate the masses of particles are zero. The interparticle interactions are also zero. Radiation pressure is zero as well. See section 15.1 and Fig. 15.1.1 in particular. Thus only Y-field black body radiation causes expansion. Its energy is calculated in this section.

$$\cong 2.9 \times 10^{194} \ \text{gm/cm}^3 = 1.63 \times 10^{218} \ \text{GeV/cm}^3$$

$$\rho_{\text{BBRW}} (t = 0, \v{r}=1) \cong (3/8\pi G)[2H_0\Omega_\gamma^{1/2}/\gamma^2 + k] \cong (3/4\pi\gamma^2 G)H_0\Omega_\gamma^{1/2} \qquad (13.7.4)$$
$$\cong 1.45 \times 10^{194} \ \text{gm/cm}^3 = 8.16 \times 10^{217} \ \text{GeV/cm}^3$$

$$\rho_{\text{BBRW}} (t = t_c, \v{r} = 0) \cong 3/(8\pi G)[H_0\Omega_\gamma^{1/2}/\gamma^2 + k] \cong 3/(8\pi\gamma^2 G)H_0\Omega_\gamma^{1/2} \qquad (13.7.5)$$
$$\cong 7.26 \times 10^{193} \ \text{gm/cm}^3 = 4.08 \times 10^{217} \ \text{GeV/cm}^3$$

$$\rho_{\text{BBRW}} (t = t_c, \v{r}=1) \cong 3/(8\pi G)[H_0\Omega_\gamma^{1/2}/(2\gamma^2) + k] \cong 3/(16\pi\gamma^2 G)H_0\Omega_\gamma^{1/2} \qquad (13.7.6)$$
$$\cong 3.63 \times 10^{193} \ \text{gm/cm}^3 = 2.04 \times 10^{217} \ \text{GeV/cm}^3$$

At t = 0 we estimate the total energy using the average:

$$\rho_{\text{BBRWaverage}} (t = 0) = (\rho_{\text{BBRW}} (t = 0, \v{r} = 0) + \rho_{\text{BBRW}} (t = 0, \v{r}=1))/2$$
$$= 12.23 \times 10^{217} \ \text{GeV/cm}^3 \qquad (15.3.4)$$

and the volume from eq. 15.3.2

$$V(0) = 4.37 \times 10^{-192} \ \text{cm}^3 \qquad (15.3.5)$$

to obtain the total Y-field expansion energy

$$E_{\text{BB}}(0) = 5.34 \times 10^{35} \ \text{eV} \qquad (15.3.6)$$

At t = t_c we estimate the total energy using the average:

$$\rho_{\text{BBRWaverage}} (t = t_c) = (\rho_{\text{BBRW}} (t = t_c, \v{r} = 0) + \rho_{\text{BBRW}} (t = t_c, \v{r}=1))/2$$
$$= 3.06 \times 10^{217} \ \text{GeV/cm}^3 \qquad (15.3.7)$$

and the volume from eq. 15.3.3

$$V(t_c) = 2.64 \times 10^{-191} \ \text{cm}^3 \qquad (15.3.8)$$

to obtain the total Y-field expansion energy

$$E_{BB}(t_c) = 8.07 \times 10^{35} \text{ eV} \qquad (15.3.9)$$

The difference between eqs. 15.3.6 and 15.3.9 can be attributed to round off error associated with averaging the densities.

15.5 A Quantum Big Bang and Evolutionary Theory

At this point we have shown that our Quantum Big Bang theory does not have a singularity at the beginning of the universe. It exhibits the known behavior of the universe since 380,000 years after the Big Bang epoch. Thus the long sought inflaton field turns out to be the quantum field, $Y^\mu(y)$, appearing in our definition of quantum coordinates. Remarkably our quantum coordinates also eliminate the divergences that have plagued quantum field theory for almost eighty years and enable calculations in The Extended Standard Model and Quantum Gravity to be divergence free without the cumbersome renormalization techniques that were needed for ElectroWeak theory and would not work for Quantum Gravity.

So our quantum coordinates free The Extended Standard Model and Quantum Gravity of infinities, both now, and at the beginning of the universe. Naturally this happy removal of infinities through one simple mechanism is a remarkable result that, we believe, reflects the simplicity of Nature when properly understood.Leibniz's Minimax Principle is realized by Two-Tier quantum theory, which removes the infinities in The Extended Standard Model, Quantum Gravity, and at the beginning of the universe.

16. Summary of the Influx-Less Epochs Case

The Expansion of the universe has a Radiation-dominated epoch and a Matter-dominated epoch. In this chapter we will summarize relevant quantities in these epochs as well as the Explosive and Big Bang epochs *assuming* that there is no influx of mass-energy from the Megaverse. Fig. 16.1 displays the data in a table format.

16.1 The Four Epochs

T	Time	Re a(t)	Im a(t)	a(t, ř)
t_{now}	13.8 Gyr	1	-6.38×10^{-93}	
Explosive			↑	$\exp(H_0\Omega_\Lambda^{1/2} t)$
t_E	9.1 Gyr	,76		
Matter				$2H_0\Omega_m^{1/2} t)^{2/3}$
t_{RM}	2.2×10^{-5} Gyr	1.8×10^{-93}		
Radiation				$(2H_0\Omega_\gamma^{1/2} t)^{1/2}$
t_c	1.26×10^{-165} s	6.38×10^{-93}		
Big Bang metastate				Re $a_{BBRW}(t, ř) = (\gamma + a(t))/2$ Im $a_{BBRW}(t, ř) = (\gamma + a(t)) ř/2$
0	0 s	3.19×10^{-93}	-3.19×10^{-93}	
?				

Figure 16.1 Times and scale factors of the four epochs.

16.2 Big Bang Metastate Data Summary

Fig. 16.2 summarizes the results for the Big Bang metastate and the radiation-dominated–matter-dominated boundary t_{RM}.

Time	0	t_c
Phase	Big Bang Beginning	Metastable End
Time	0	1.26×10^{-165}
Re a(t)	8.16×10^{-93}	1.632×10^{-92}
Im a(t)	-3.16×10^{-93}	-5.24×10^{-93}
Re radius	4.278×10^{-65} cm	8.5×10^{-65} cm
Im radius	4.278×10^{-65} cm	8.5×10^{-65} cm
Volume	4.37×10^{-192} cm^3	2.6398×10^{-191} cm^3
Central Expansion Energy Density	1.63×10^{218} GeV/cm^3	4.08×10^{217} GeV/cm^3
Edge Expansion Energy Density	8.16×10^{217} GeV/cm^3	2.04×10^{217} GeV/cm^3
Total Expansion Energy[87]	5.34×10^{35} eV	8.07×10^{35} eV
Hubble Constant[88] (km s^{-1} Mpc^{-1})	1.79×10^{218}	1.14×10^{126}

Figure 16.2 The Big Bang Epoch detailed data.

---

[87] The expansion energy does not include the mass-energy of particles in the Big Bang universe. It only includes y field black body energy which drives the initial expansion. All particles are massless initially and all interparticle forces are zero.

[88] In the Big Bang metastable state we display the Hubble constant at the "expanding' edge.

16.3 Some Background Equations for Big Bang Metastate Data

Below are some of the equations supporting the data in Fig. 16.2.

$\check{r} \in [0, 1]$ From center to edge of Big Bang Metastable universe

$0 \le t \le t_c$

$$\text{Re } a_{BBRW}(t, \check{r}) \cong \gamma/2 + a(t)/2 \qquad (13.3.2.24)$$

$$\text{Im } a_{BBRW}(t, \check{r}) \cong -\gamma\check{r}/2 - a(t)\check{r}/2 \qquad (13.3.2.25)$$

$t_c < t$

$$\text{Re } a_{BBRW}(t, \check{r}) \cong a(t)\{1 + (1 + \check{r}^2)\gamma^2/4\} \cong a(t) \qquad (13.3.2.26)$$

$$\text{Im } a_{BBRW}(t, \check{r}) \cong -\gamma\check{r}/2 - [\check{r}^2 + \gamma^2/(4a^2(t))]^{\frac{1}{2}}\gamma/2 \qquad (13.3.2.27)$$

$t \to \infty$

$$\text{Re } a_{BBRW}(t, \check{r}) \cong a(t)\{1 + (1 + \check{r}^2)\gamma^2/4\} \cong a(t) \qquad (13.3.2.28)$$

$$\text{Im } a_{BBRW}(t, \check{r}) \cong -\gamma\check{r} = -8.166 \times 10^{-93}\check{r} \qquad (13.3.2.29)$$

$0 \le t < t_c$

$$\kappa T_< \equiv \kappa T(0 \le t < t_c) \cong 2\kappa T_0(1 + i\check{r})/\chi = 4\pi^{-3/2} M_c(1 + i\check{r})$$
$$= 2.72 \times 10^{29}(1 + i\check{r}) \text{ eV} \qquad (13.3.3.4)$$
$$= 3.155 \times 10^{33}(1 + i\check{r}) \,^\circ K$$

$t_c < t$

$$\kappa T_> \equiv \kappa T(t_c < t)$$
$$\cong \kappa T_0(a(t) + i\gamma\check{r})/a^2(t) = \kappa T_0/a(t) \qquad (13.3.3.5)$$

$0 \le t < t_c$

$$H(t, \check{r}) = H_{BBRW}(t, \check{r}) \cong 4H_0\Omega_\gamma^{\frac{1}{2}}|[1 - \check{r}^2 + 2i\check{r}]/[(\gamma + a(t))^2(1 + \check{r}^2)^2]| \qquad (13.4.5)$$

$$H_{BBRW}(0, \check{r}) \cong 4H_0\Omega_\gamma^{\frac{1}{2}}|[1 - \check{r}^2 + 2i\check{r}]/[\gamma^2(1 + \check{r}^2)^2]| \qquad (13.4.6)$$

$$H_{BBRW}(0, 1) \cong 2H_0\Omega_\gamma^{\frac{1}{2}}/\gamma^2 = 5.8 \times 10^{198} \text{ s}^{-1} \qquad (13.4.7)$$
$$= 1.79*10^{218} \text{ km s}^{-1} \text{ Mpc}^{-1}$$

$$H_{BBRW}(0, 0) \cong 4H_0\Omega_\gamma^{\frac{1}{2}}1/\gamma^2 = 2H_{BBRW}(0, 1) \qquad (13.4.8)$$

$$H_{BBRW}(t_c, 1) \cong 4H_0\Omega_\gamma^{\frac{1}{2}}|i|/[2(\gamma + a(t_c))^2]| \qquad (13.4.9)$$
$$= 2\,H_0\Omega_\gamma^{\frac{1}{2}}/(\gamma + \gamma^2) \cong 2\,H_0\Omega_\gamma^{\frac{1}{2}}/\gamma = 3.7 \times 10^{106}\ s^{-1}$$
$$= 1.14*10^{126}\ km\ s^{-1}\ Mpc^{-1}$$

$$H_{BBRW}(t_c, 0) \cong H_0\Omega_\gamma^{\frac{1}{2}}/\gamma^2 = 2.9 \times 10^{198}\ s^{-1} \qquad (13.4.10)$$
$$= 8.95*10^{217}\ km\ s^{-1}\ Mpc^{-1}$$

$$r_{universeBBRW}(t) = \{-i\gamma + a(t) + [\gamma^2 - 2i\gamma a(t) + a^2(t)]^{\frac{1}{2}}\}/(2k^{\frac{1}{2}}) \qquad (15.2.3)$$

$$Re\ r_{universeBBRW}(t) = \{a(t) + [R(1 + \cos\psi)/2]^{\frac{1}{2}}\}/(2k^{\frac{1}{2}}) \qquad (15.2.4a)$$

$$Im\ r_{universeBBRW}(t) = \{-\gamma - [R(1 - \cos\psi)/2]^{\frac{1}{2}}\}/(2k^{\frac{1}{2}}) \qquad (15.2.4b)$$

$$R = [(\gamma^2 + a^2(t))^2 + 4\gamma^2 a^2(t)]^{\frac{1}{2}} \qquad (15.3.2.12)$$
$$\cos\psi = (\gamma^2 + a^2(t))/R \qquad (15.3.2.13)$$

$$Re\ r_{universeBBRW}(0 \le t < t_c) \cong (\gamma + a(t))/(2k^{\frac{1}{2}}) \qquad (15.2.5a)$$
$$Re\ r_{universeBBRW}(0) \cong \gamma/(2k^{\frac{1}{2}}) = 4.278 \times 10^{-65}\ cm \qquad (15.2.5b)$$
$$Im\ r_{universeBBRW}(0 \le t < t_c) \cong -(\gamma + a(t))/(2k^{\frac{1}{2}}) \qquad (15.2.6a)$$
$$Im\ r_{universeBBRW}(0) \cong -\gamma/(2k^{\frac{1}{2}}) = -4.278 \times 10^{-65}\ cm \qquad (15.2.6b)$$

$$r_{universeBBRW}(t_{now}) = a(t_{now})/k^{\frac{1}{2}} - i\gamma/k^{\frac{1}{2}} \qquad (15.2.7)$$
$$Re\ r_{universeBBRW}(t_{now}) = a(t_{now})/k^{\frac{1}{2}} = 1.34 \times 10^{28}\ cm \qquad (15.2.8)$$
$$Im\ r_{universeBBRW}(t_{now}) = -\gamma/k^{\frac{1}{2}} = -8.55 \times 10^{-65}\ cm \qquad (15.2.9)$$

$$\rho_{BBRW}\,(t = 0, \check{r} = 0) \cong (3/8\pi G)[4H_0\Omega_\gamma^{\frac{1}{2}}/\gamma^2 + k] \cong (3/2\pi\gamma^2 G)H_0\Omega_\gamma^{\frac{1}{2}} \qquad (13.7.3)$$
$$\cong 2.9 \times 10^{194}\ gm/cm^3 = 1.63 \times 10^{218}\ GeV/cm^3$$

$$\rho_{BBRW}\,(t = 0, \check{r}=1) \cong (3/8\pi G)[2H_0\Omega_\gamma^{\frac{1}{2}}/\gamma^2 + k] \cong (3/4\pi\gamma^2 G)H_0\Omega_\gamma^{\frac{1}{2}} \qquad (13.7.4)$$
$$\cong 1.45 \times 10^{194}\ gm/cm^3 = 8.16 \times 10^{217}\ GeV/cm^3$$

$$\rho_{BBRW} (t = t_c, \check{r} = 0) \cong 3/(8\pi G)[H_0\Omega_{,\gamma}^{1/2}/\gamma^2 + k] \cong 3/(8\pi\gamma^2 G)H_0\Omega_{,\gamma}^{1/2} \quad (13.7.5)$$
$$\cong 7.26 \times 10^{193} \ gm/cm^3 = \ 4.08 \times 10^{217} \ GeV/cm^3$$

$$\rho_{BBRW} (t = t_c, \check{r}=1) \cong 3/(8\pi G)[H_0\Omega_{,\gamma}^{1/2}/(2\gamma^2) + k] \cong 3/(16\pi\gamma^2 G)H_0\Omega_{,\gamma}^{1/2} \quad (13.7.6)$$
$$\cong 3.63 \times 10^{193} \ gm/cm^3 = \ 2.04 \times 10^{217} \ GeV/cm^3$$

Reader Note

Chapters 17 – 20 describe a calculation of the growth of the universe using the Einstein equation and assuming an influx of energy from the Megaverse. For reasons given later we will see that this approach fails due to additional large energy density and pressure. Nevertheless it appears appropriate to present this description to illustrate this method of calculation.

The reader may skip to chapter 21 for a more physically acceptable calculation of the evolution of the universe from the Big Bang Metastate to the present.

17. Conventional Universe Expansion with an Influx from the Megaverse

This chapter calculates the mass-energy influx from the Megaverse and the consequent change in energy density with time. As a result of the influx, the character of the epochs changes and the Hubble Constant becomes time dependent as current experiment seems to indicate.

17.1 Initial State for Influx Initiation

If we examine the time periods of the universe epochs in the absence of a Megaverse influx, then it appears that the beginning of the influx can be set to a time within the Matter-dominated epoch.[89]

Because available data[90] gives Hubble Constant values for the early universe at time $t = t_T = 380,000$ years $= 1.19837 \times 10^{13}$ s $= 1.205 \times 10^{-11}$ Gyr, a time within the Matter-dominated epoch we will calculate the change in the universe since that time due to the influx between then and now. Note that the 380,000 years before t_T is negligible compared to the 13.8 Gyr existence of the universe. Consequently the influx before t_T will be negligible compared to the influx in the 13.4 Gyr period afterwards.

We find the scale factor at $t_T = 380,000$ years to be

$$a(t_T) = (2H_0\Omega_\gamma^{1/2} t_T)^{1/2} \qquad (17.1)$$
$$= 0.0066 \qquad (17.2)$$

---

[89] See Fig. 16.1.
[90] See, for example, CMB data of K. Aylor *et al*, arXiv:1811.00537v1 (2018) and similar studies.

17.2 Megaverse Influx Equation for the Energy Density

In chapter 14 we determined the time-dependent equation for the Megaverse influx:

$$d\Omega_T(t)/dt = C_M k^{1/2}[\Omega_H(t) + \Omega_T(t)]/a(t) \qquad (14.4)$$

The solution to eq. 14.4 is[91]

$$\Omega_T(t) = C_M k^{1/2} e^{F(t)} \int_{t_T}^{t} dt'' \, e^{-F(t'')} \Omega_H(t'')/a(t'') \qquad (14.5)$$

where

$$F(t) = C_M k^{-1/2} \int_{t_T}^{t} dt'/a(t') \qquad (14.2)$$

17.3 Approximate Solution Based on Matter-Dominated a(t)

We can approximate eq.14.4 based on the assumption that $\Omega_H(t) \gg \Omega_T(t)$. This assumption seems reasonable since the increasing expansion rate of the universe seems relatively modest. We attribute the "modestly" increasing rate to a modest mass-energy influx from the Megaverse. Its relative "smallness" is due to the dominance of $\Omega_H(t)$ over the incremental mass-energy factor $\Omega_T(t)$. Therefore we approximate

$$d\Omega_T(t)/dt = C_M k^{1/2}\Omega_H(t)/a(t) \qquad (17.3)$$

Eq. 17.3 can be solved:[92]

$$\Omega_T(t) = C_M k^{1/2} \int_{t_T}^{t} dt'\Omega_H(t')/a(t') \qquad (17.4)$$

---

[91] This approximation causes the energy influx to not "feed" on itself after it enters our universe. Thus, to some extent, it limits the growth of the energy density and the Hubble Constant..

[92] This approximation yields a lower value for the total energy density. As a result da/dt will be smaller and the Hubble constant obtained will be lower.

Combining eq. 17.4 with the Einstein equation:

$$\dot{a}^2(t) - 8\pi G \rho_{tot} a^2(t)/3 = -k \qquad (11.19)$$

we have an implicit equation for a(t) in the Megaverse mass-energy influx epoch, which continues up to the present day:

$$\dot{a}^2(t) - (8\pi G/3)\, a^2(t)\rho_{crit}(\Omega_H(t) + \Omega_T(t)) = -k \qquad (17.5)$$

where

$$\Omega_H(t) = \Omega_\gamma(t) + \Omega_m(t) + \Omega_d(t) + \Omega_\Lambda \qquad (9.10)$$

with

$$\Omega_\gamma(t) = 5.38 \times 10^{-5}/a^4(t) \qquad (17.6)$$
$$\Omega_m(t) = 0.046/a^3(t)$$
$$\Omega_d(t) = 0.268/a^3(t)$$
$$\Omega_\Lambda = 0.692$$
$$\Omega_M(t) = \Omega_m(t) + \Omega_d(t) = 0.308/a^3(t) \qquad (17.7)$$

The dominant terms in $\Omega_H(t)$ are the sum of the baryon density, the cold dark matter density, and the dark energy density:

$$\Omega_H(t) \approx \Omega_M(t) + \Omega_\Lambda = \Omega_M/a^3(t) + \Omega_\Lambda \qquad (17.8)$$

where

$$\Omega_M = 0.308 \qquad (17.8a)$$

Thus

$$\Omega_T(t) = C_M k^{\frac{1}{2}} \int_{t_T}^{t} dt'[\Omega_M/a^4(t') + \Omega_\Lambda/a(t')] \qquad (17.9)$$

Combining the above equations we obtain

$$da/dt \approx H_0 \left[\Omega_M /a(t) + \Omega_\Lambda a^2(t) + C_M k^{\frac{1}{2}} a^2(t) \int_{t_T}^{t} dt' \, (\Omega_M/a^4(t') + \Omega_\Lambda/a(t'))\right]^{\frac{1}{2}} \qquad (17.10)$$

To calculate a(t) we will assume that it has the same form as the Matter dominated epoch in chapter 13 since we use $t_T = 380,000$ years[93] to allow direct comparison with astrophysical Hubble Constant data:

:

$$a(t) = \beta t^\gamma \qquad (17.11)$$

with the normalization

$$a(t_{now}) = 1 \qquad (17.12)$$

resulting in

$$\beta = t_{now}^{-\gamma} \qquad (17.12a)$$

Then after a time integration eq. 17.10 becomes

$$a(t) - a(t_T) \approx H_0 \int_{t_T}^{t} dt' \, [\Omega_M/(\beta t'^\gamma) + \Omega_\Lambda(\beta t'^\gamma)^2 + d_M[\Omega_M \beta^{-2}(t'^{1-2\gamma} - t'^{2\gamma} t_T^{1-4\gamma})/(1-4\gamma) +$$

$$+ \Omega_\Lambda \beta(t'^{1+\gamma} - t_T t'^{1-\gamma} t'^{2\gamma}/(1-\gamma))]^{\frac{1}{2}} \qquad (17.13)$$

where

$$d_M = C_M k^{\frac{1}{2}} \qquad (17.14)$$

Note eq. 17.11 and 17.13 imply

$$\Omega_T(t) = d_M [\Omega_M \beta^{-4}(t^{1-4\gamma} - t_T^{1-4\gamma})/(1-4\gamma) + \Omega_\Lambda(t^{1-\gamma} - t_T^{1-\gamma})/(\beta(1-\gamma))] \qquad (17.9a)$$

Since the second and following terms on the right (the influx increment) in eq. 17.13 are presumably[94] small compared to the first term we approximate eq. 17.13 with

---

[93] This time is within the matter-dominated epoch using the conventional a(t) of chapter 13. Note that the 380,000 years before $t = t_{cSE}$ is negligible compared to the 13.8 Gyr existence of the universe. Cosequently the influx before t_{cSE} will be negligible compared to the influx in the 13.4 Gyr period afterwards.

[94] We check this in section 17.5.

$$a(t) - a(t_T) \approx H_0 \int_{t_T}^{t} dt' \sqrt{\Omega_M} \, (\beta t'^{\gamma})^{-\frac{1}{2}} \, [1 + \frac{1}{2} \, (\Omega_\Lambda/\Omega_M)(\beta t'^{\gamma})^3 + \frac{1}{2}(\beta t'^{\gamma}/\Omega_M) \, \Omega_T] \tag{17.15}$$

$$= H_0 \int dt' [\sqrt{\Omega_M} \, (\beta t'^{\gamma})^{-\frac{1}{2}} + \frac{1}{2} \sqrt{\Omega_M} \, (\beta t'^{\gamma})^{-\frac{1}{2}} \, (\Omega_\Lambda/\Omega_M)(\beta t'^{\gamma})^3 + \frac{1}{2}\Omega_M^{-\frac{1}{2}}\beta^{\frac{1}{2}}t'^{\gamma/2} \, d_M \cdot$$
$$\cdot [\Omega_M \beta^{-4}(t^{1-4\gamma} - t_T^{1-4\gamma})/(1-4\gamma) + \Omega_\Lambda(t^{1-\gamma} - t_T^{1-\gamma})/(\beta(1-\gamma))]]$$

$$= H_0 \, \beta^{-\frac{1}{2}}\Omega_M^{1/2} \{(t^{1-\gamma/2} - t_T^{1-\gamma/2})/(1-\gamma/2) +$$
$$+ \frac{1}{2} \, \beta^3(\Omega_\Lambda/\Omega_M)(t^{2.5\gamma+1} - t_T^{2.5\gamma+1})/(2.5\gamma+1) +$$
$$+ \frac{1}{2}\beta\Omega_M^{-1}d_M\{\Omega_M\beta^{-4}(t^{2-3.5\gamma}/(2-3.5\gamma) -$$
$$- t_T^{1-4\gamma} \, t^{\gamma/2+1}/(1+\gamma/2))/(1-4\gamma) +$$
$$+ \Omega_\Lambda(t^{-\gamma/2+2}/(2-\gamma/2) - t_T^{1-\gamma} \, t^{\gamma/2+1}/(1+\gamma/2))/(\beta(1-\gamma)) -$$

$$- \frac{1}{2}\beta\Omega_M^{-1}d_M\{\Omega_M\beta^{-4}(t_T^{2-3.5\gamma} \, (1/(2-3.5\gamma) - 1/(1+\gamma/2))/(1-4\gamma) +$$
$$+ \Omega_\Lambda(t_T^{-\gamma/2+2}(1/(2-\gamma/2) - 1/(1+\gamma/2))/(\beta(1-\gamma))\}$$

with $a(t_T) = 0.0066$ by eq. 17.2 where $t_T = 380,000$ years $= 1.19837 \times 10^{13}$ s.

Upon integration we obtain

$$a(t) - a(t_T) \approx H_0 \beta^{-\frac{1}{2}} \sqrt{\Omega_M} \, (V(t) - V(t_T)) \tag{17.16}$$

where $a(t_T)$ is the matter-dominated scale factor value at time t_T and

$$V(t) = t^{1-\gamma/2}/(1-\gamma/2) + \frac{1}{2}\beta^3(\Omega_\Lambda/\Omega_M)t^{2.5\gamma+1}/(2.5\gamma+1) + \tag{17.17}$$
$$+ \frac{1}{2} \, d_M\{\beta^{-3} \, (t^{2-3.5\gamma}/(2-3.5\gamma) - t_T^{1-4\gamma} \, t^{\gamma/2+1}/(1+\gamma/2))/(1-4\gamma) +$$
$$+ (\Omega_\Lambda/\Omega_M)(t^{-\gamma/2+2}/(2-\gamma/2) - t_T^{1-\gamma} \, t^{1+\gamma/2}/(1+\gamma/2))/(1-\gamma)\}$$

The derivative of $a(t)$ is

$$da(t)/dt \approx H_0\beta^{-\frac{1}{2}} \sqrt{\Omega_M} \, dV(t)/dt \tag{17.16a}$$

Since we assume that the influx density is much smaller than the matter energy density in the matter-dominated phase during which the influx occurs we will use the matter epoch a(t) where $\gamma = 2/3$ and $\beta = t_{now}^{-\gamma}$, and treat the influx contribution as based on the total matter energy density. Then we see that V(t) becomes[95]

$$V_M(t) = 3t^{2/3}/2 + (3/16)\, t_{now}^{-2}(\Omega_\Lambda/\Omega_M)t^{8/3} + \quad\quad (17.18)$$
$$+(9/2)\, d_M\{t_{now}^2(t^{-1/3} + t_T^{-5/3}\, t^{4/3}/4)/5 +$$
$$+ (\Omega_\Lambda/\Omega_M)(t^{5/3}/5 - t_T^{1/3}\, t^{4/3}/4)\}$$

and

$$V_M(t_T) = 3t^{2/3}/2 + (3/16)\, t_{now}^{-2}(\Omega_\Lambda/\Omega_M)t^{8/3} + \quad\quad (17.18a)$$
$$+(9/8)\, d_M\{t_{now}^2\, t_T^{-1/3} - (1/20)\,(\Omega_\Lambda/\Omega_M)\, t_T^{5/3}\}$$

Note the first term $t^{2/3}$ leading dependence of a(t) as expected from the choice of γ. We will see that the influx terms of eq. 17.18 causes a(t) and the Hubble parameter ("constant") to increase with time as seems to be the findings of astrophysical studies.

We define

$$a_M(t) = H_0\beta^{-\frac{1}{2}} \sqrt{\Omega_M}\, (V_M(t) - V_M(t_T)) + a(t_T) \quad\quad (17.19)$$

where $a(t_T)$ is the matter-dominated scale factor value at time t_T and the derivative

$$da_M(t)/dt \approx H_0\beta^{-\frac{1}{2}} \sqrt{\Omega_M}\, dV_M(t)/dt \qu\quad (17.19a)$$

where

$$dV_M(t)/dt = t^{-1/3} + \frac{1}{2}\, t_{now}^{-2}(\Omega_\Lambda/\Omega_M)t^{5/3} + \quad\quad (17.20)$$
$$+(3/2)d_M\{t_{now}^2(-t^{-4/3} + t_T^{-5/3}\, t^{1/3})/5 + (\Omega_\Lambda/\Omega_M)(t^{2/3} - t_T^{1/3}\, t^{1/3})\}$$

---

[95] The coefficient of the $t^{2/3}$ term in a(t) expression inf eqs. 17.16 and 17.18 differs from the standard term:

$$a(t) \cong [2H_0\Omega_m^{\frac{1}{2}}t]^{2/3} \qu\quad (13.2.3.8)$$

by a factor of 1.19 more due to the difference between the algebraic approximation giving eq. 13.2.3.8 and the above integration. This difference "washes out" when the Hubble Constant is calculated H = d(ln a)/dt.

We will use it in section 17.6 to determine the time dependence of the Hubble parameter (Constant) $H_M(t)$:

17.4 Evaluation of d_M and the Determination of C_T and ρ_{Mega}

The parameter d_M above can be expressed in terms of the universe's surface tension and G using eqs. 9.8 and 17.14 which give

$$d_M = C_T k^{\frac{1}{2}} G \rho_{Mega} \qquad (17.21)$$

Since d_M must have the dimension s^{-1} by eq. 17.18, we define

$$C_T = C_{T1} \, ck^{-1} \qquad (17.21a)$$

making C_{T1} a dimensionless constant where c is the speed of light. Then

$$\begin{aligned} d_M &= C_{T1} \, ck^{-1/2} G \rho_{Mega} \\ &= C_{T1} \, ck^{-1/2} G \rho_{crit} \rho_{M0} \end{aligned} \qquad (17.22)$$

where C_{T1}, and ρ_{M0} are dimensionless constants, with

$$\rho_{Mega} = \rho_{M0} \, \rho_{crit} \qquad (17.23)$$

We express ρ_{Mega} in terms of the multiple ρ_{M0} of our universe's critical density ρ_{crit}. C_{T1} is a surface tension constant which we will set to one in this chapter and the following chapters[96]

$$C_{T1} = 1 \qquad (17.24)$$

---

[96] The value of C_{T1} probably should have a different constant value. When we later fit the model to the data to compare with experiment only the value of $C_{T1} \rho_{M0}$ will be fixed. We will suggest $C_{T1} = (L_{Planck} k^{\frac{1}{2}})^{-1} \approx 10^{62}$ where L_{Planck} is the Planck length since we will find $C_{T1} \rho_{M0}$ is of the order of 10^{-8}. Thus the Megaverse density factor ρ_{M0} will be of the order of 10^{-70}. (A physically reasonable value since a universe is a much higher density region of the Megaverse.) The choice of the dimensionless constant $C_{T1} = (L_{Planck} k^{\frac{1}{2}})^{-1}$ is reasonable since both factors in it refer to the curvature of the universe. Surface Tension depends on the radius of curvature.

Eq. 17.24 implies

$$d_M = 2.57 \times 10^{-19} \text{ s}^{-1} \, \rho_{M0} \qquad (17.25)$$

17.5 Megaverse Energy Density Bound

Since $\Omega_M = 0.308$ and $\Omega_\Lambda = 0.692$, we expect $\Omega_T(t_{now})$ to be of the order of 1 or less giving an *upper* bound on ρ_{M0} of 10^{-6}. The Megaverse density, $\rho_{Mega} = \rho_{M0} \, \rho_{crit}$, is thus very much less than our universe's critical energy density ρ_{crit}. If the Megaverse measure of density $\Omega_T(t_{now})$ is of the order of magnitude of Ω_M, then we expect $\Omega_T(t_{now})$ to be between $0 - 1$, with the result the order of magnitude of ρ_{M0} is between 0 and 10^{-6} based on eq. 17.24. Since we wish to keep $\Omega_T(t_{now}) < \Omega_M$ for physical reasons, we will examine values of $C_{T1} \, \rho_{M0}$ below 10^{-6} in chapter 18.

17.6 The Time Dependent Hubble Constant H(t)

The time dependent Hubble "Constant" can be expressed in terms of $V_M(t)$ and its derivative as

$$H_M(t) = da_M(t)/dt \,/\, a_M(t) = dV_M(t)/dt \,/[\, V_M(t) - V_M(t_T) + (\, H_0 \beta^{-\frac{1}{2}} \sqrt{\Omega_M})^{-1} a(t_T)] \qquad (17.26)$$

using eqs. 17.18 – 17.20. In chapter 20 we will use an alternate, more convenient, form for H(t):

$$H(t) = (da/dt)/a(t) = [H_0{}^2 \rho_{tot}(t)/\rho_{cri} - c^2 k/a^2(t)]^{\frac{1}{2}} \qquad (20.1)$$

18. "Conventional" Influx Megaverse Mass-Energy Accretion in our Universe

In this chapter we determine the time dependence of the Megaverse mass-energy influx as a function of time and its impact on the density of the universe.

18.1 Influx Mass-Energy Density $\rho_T(t)$ as a Function of Time

Eq. 17.9a implies the influx mass-energy density from the Megaverse is

$$\rho_T(t) = \rho_{crit}\Omega_T(t) \tag{18.1}$$

where

$$\Omega_T(t) = d_M \left[\Omega_M\beta^{-4}(t^{1-4\gamma} - t_T^{1-4\gamma})/(1-4\gamma) + \Omega_\Lambda(t^{1-\gamma} - t_T^{1-\gamma})/(\beta(1-\gamma))\right] \tag{17.9a}$$

Eq. 18.1 is plotted in Fig. 18.1. Note $\Omega_T(t)$ is approximately constant. The values of ρ_{M0} and the average value of $\Omega_T(t)$ are

ρ_{M0}	Average $\Omega_T(t)$
3.7037E-08	0.0304866
7.40741E-08	0.0609732
1.48148E-07	0.1219463
2.96296E-07	0.2438927
5.92593E-07	0.4877854
1.18519E-06	0.9755708

Table 18.1 Megaverse Density Factors and average $\Omega_T(t)$. The corresponding graph values in the following figures, in descending order, are 2, 20, 40, 60, 80, and 100.

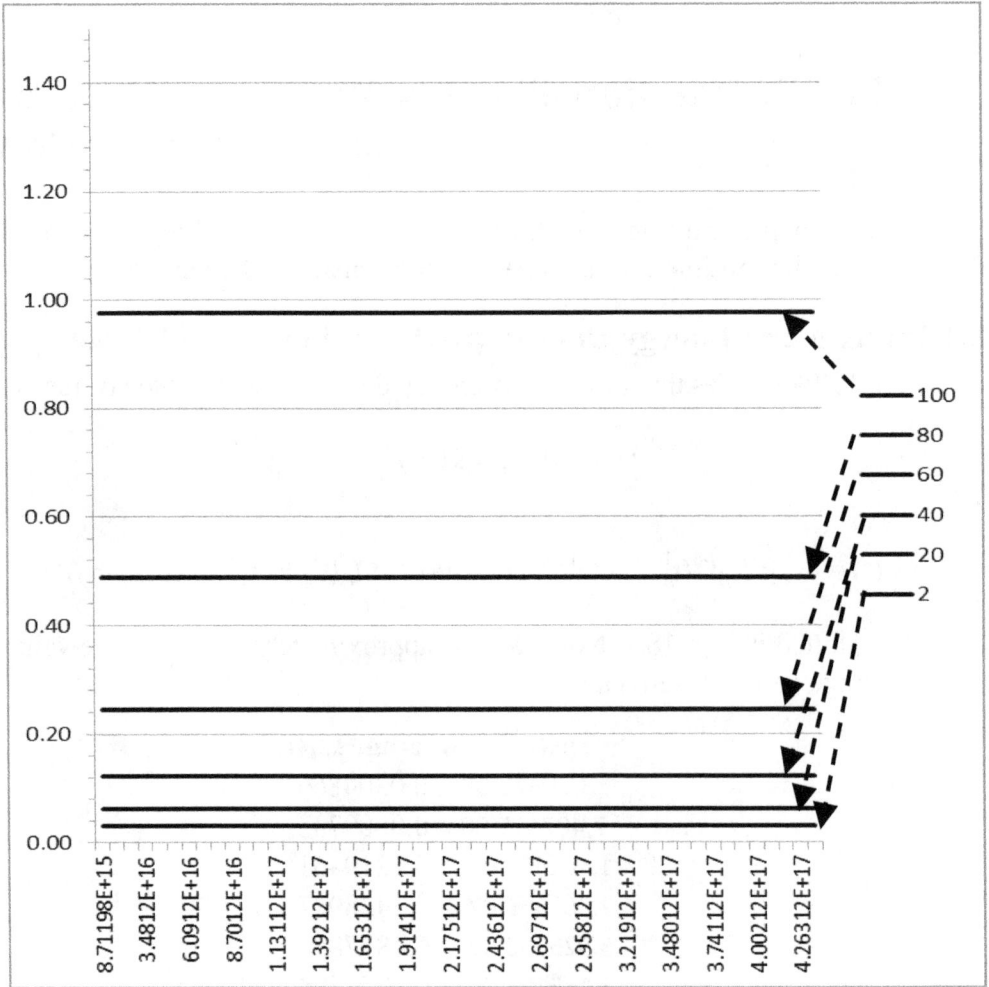

Figure 18.1 Plots of $\Omega_T(t)$ vs. time in seconds where Megaverse mass-energy density $\rho_{Mega} = \rho_{M0}\rho_{crit}$. The number and line ordering corresponds to the ordering in Table 18.1.

Note: If C_{T1} is very large than the Megaverse density ρ_{M0} factor will be very small. .

Note also, that the flatness of the curves in Fig. 18.1 suggests that we can use the average values given above as an approximation to $\Omega_T(t)$ in the Einstein equation for the various possible values of ρ_{M0}.

19. "Conventional" Universe Scale Factor for a Megaverse Influx Epoch

In this chapter we investigate the Influx-related scale factor $a_M(t)$ in eq. 17.19 for use in chapter 20 in the determination of the Hubble Parameter (Constant).

19.1 Influx-Driven Universe Expansion Epochs

In this section we summarize the epochs for the case of an influx-driven expansion under the assumption that it began at 380,000 years to facilitate comparison with astrophysical data. The influx, if any, that took place before that time is negligible in comparison with the influx after that time.

T	Time	Re a(t)	Im a(t)	$a(t, \check{r})$
t_{now}	13.8 Gyr	1	-6.38×10^{-93}	
Influx				$a_M(t)$
t_T	380,000 yr	$a(t_T)$		
Matter				$2H_0\Omega_m^{\frac{1}{2}} t)^{2/3}$
t_{RM}	2.2×10^{-5} Gyr	1.8×10^{-93}		
Radiation				$(2H_0\Omega_\gamma^{\frac{1}{2}} t)^{\frac{1}{2}}$
t_c	1.26×10^{-165} s	6.38×10^{-93}		
Big Bang metastate				Re $a_{BBRW}(t, \check{r}) = (\gamma + a(t))/2$ Im $a_{BBRW}(t, \check{r}) = (\gamma + a(t)) \check{r}/2$
0	0 s	3.19×10^{-93}	-3.19×10^{-93}	

Figure 19.1. The Influx Epoch and Other Epochs.

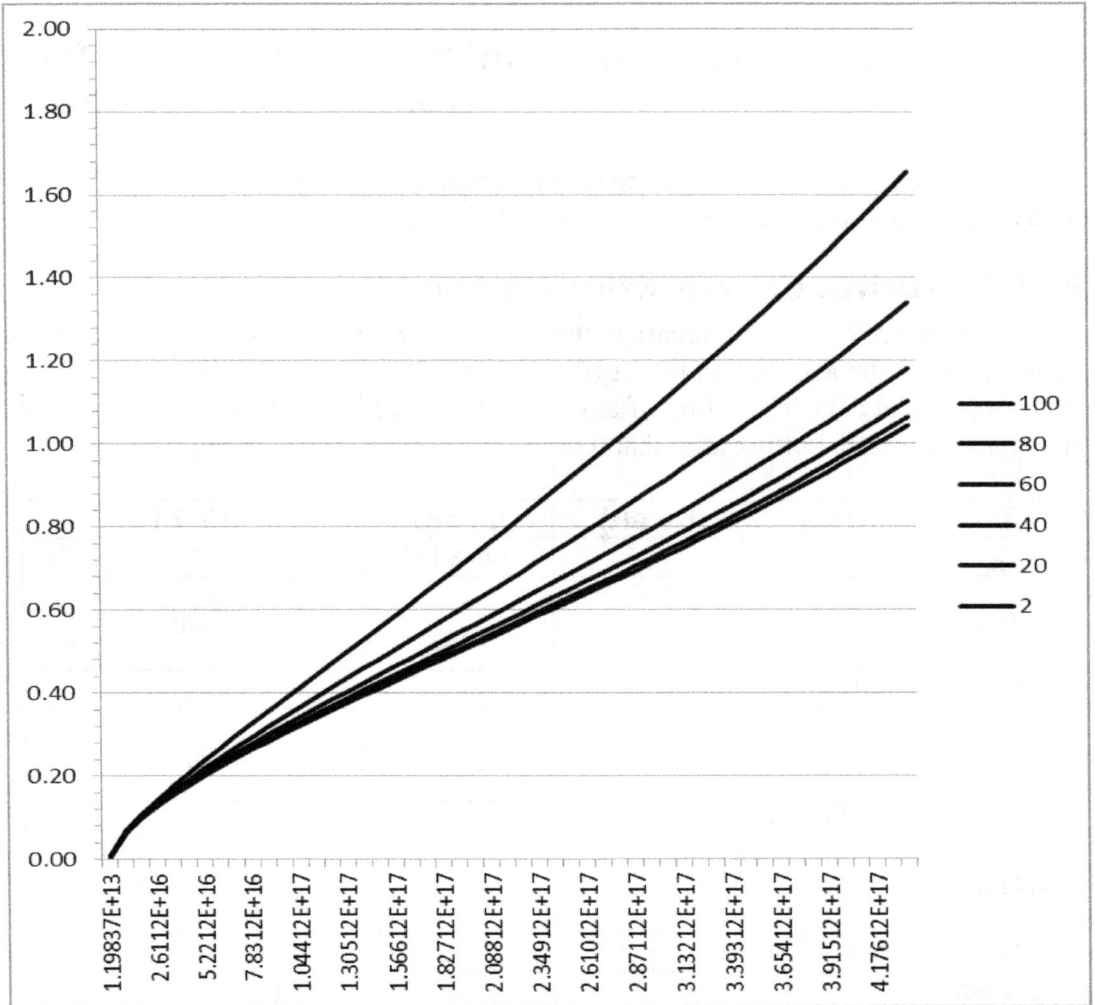

Figure 19.2. Scale factor $a_M(t)$ as a function of the time in seconds for six values of the Megaverse energy scale factor ρ_{M0}. The order of the lines is according to Table 18.1. The time ranges from 380,000 years to the present.

20. "Conventional" Hubble Constant Growth in Time

20.1 Influx Driven Hubble Constant in Time

The Einstein equation

$$\dot{a}^2(t) - 8\pi G \rho_{tot} a^2(t)/3 = -k \tag{11.19}$$

can be used to find the Hubble Parameter (Constant) as a function of time:

$$H(t) = (da/dt)/a(t) = [H_0^2 \rho_{tot}(t)/\rho_{cri} - c^2 k/a^2(t)]^{\frac{1}{2}} \tag{20.1}$$

where c is the speed of light and

$$\rho_{tot}(t) = \rho_{crit}(\Omega_H(t) + \Omega_T(t)) \tag{20.2}$$

$$\Omega_H(t) = \Omega_\gamma(t) + \Omega_m(t) + \Omega_d(t) + \Omega_\Lambda(t) \tag{9.10}$$

$$H_0^2 = 8\pi G \rho_{crit}/3 \tag{20.3}$$

Eq. 20.1 is plotted in Fig. 20.1. The negative values of H(t) squared (and the accompanying sharp edges in the curve) indicate the calculation does not capture the essence of the expansion properly.

For this reason we will examine a new approach to determining the expansion of the universe in the next chapter.

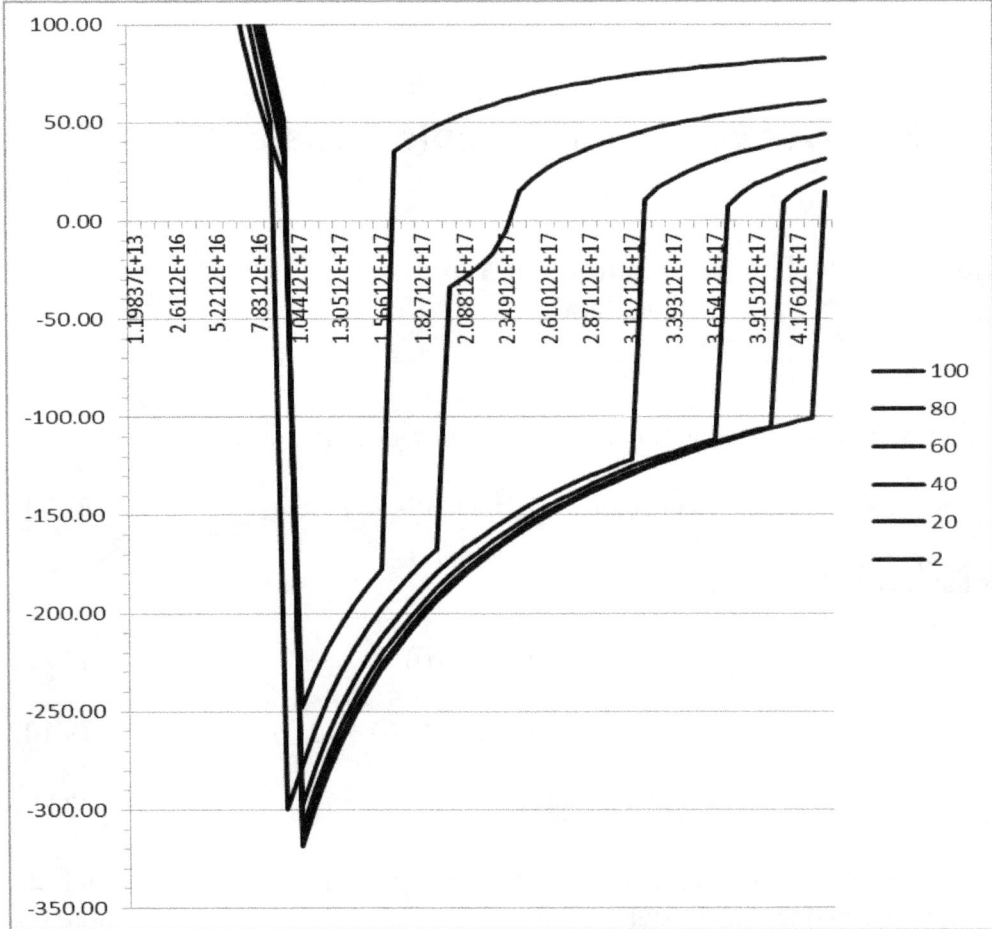

Figure 20.1. Plot of H(t) as a function of time in seconds. The apparently negative values of H(t) signal the fact that H(t) squared is negative (H is thus imaginary) at those points due to the dominance of the $c^2k/a^2(t)$ term in eq. 20.1. This figure demonstrates the inadaquacy of the approach of this section. See chapter 21 for details.

SECTION 3: UNIVERSAL SCALE FACTOR MODEL

21. The Universe Evolution Problem

The preceding chapters describe the evolution of the universe based on the Einstein equation and the standard expression for the energy density of the universe supplemented by an influx energy density from the Megaverse:

Einstein Equation

$$\ddot{a}^2 - 8\pi G \rho_{tot} a^2/3 = -k \tag{10.5.2}$$

Total Energy Density

$$\rho_{tot}(t) \equiv \rho_{crit}\Omega_{tot}(t) = \rho_{crit}[\Omega_\Gamma(t) + \Omega_M(t) + \Omega_\Lambda + \Omega_T]$$
$$\equiv \rho_{crit}\Omega_{tot}(t) = \rho_{crit}[\Omega_\Gamma(t) + \Omega_M(t) + \Omega_\Lambda + \Omega_{TST} + \Omega_{Trest}]$$

$$= \rho_{crit}[\Omega_\gamma(t) + \Omega_\Lambda + \Omega_m(t) + \Omega_d(t) + \Omega_{Trest} + C_T G \rho_{Mega} \int_{t_T}^{t} dt' \rho_{tot}(t')/R(t')] \tag{9.6}$$

where Ω_{TST} is the surface tension based Megaverse energy influx, and Ω_{Trest} is the additional Dark Energy which is set to zero in the preceding. Note

$$\Omega_T = \Omega_{TST} + \Omega_{Trest} \tag{21.1}$$

$$\Omega_H = \Omega_\gamma(t) + \Omega_\Lambda + \Omega_m(t) + \Omega_d(t) \tag{21.2}$$
$$\Omega_H = \Omega_\Gamma(t) + \Omega_\Lambda + \Omega_M(t)$$

In chapter 23 we will find Ω_T is overwhelmingly larger than Ω_H

$$\Omega_T \gg \Omega_H \tag{21.3}$$

making the calculation of chapters 17 – 20 irrelevant.

The energy conservation law is of some importance:

Energy Conservation Equation

$$d(\rho_{tot} \, a^3)/da = -3pa^2 \tag{2.2}$$

with scale factor a(t) and pressure p. The Matter and Radiation epoch scale factors are (under appropriate assumptions)

Matter-Dominated: $\qquad\qquad \rho = \rho_0/a(t)^3$ $\qquad\qquad$ (11.2)

Radiation-Dominated: $\qquad\quad \rho = \rho_0/a(t)^4$ $\qquad\qquad$ (11.3)

The calculation of the scale factor (in a reasonable approximation) in the preceding chapters yielded a Hubble Constant that was in part imaginary-valued. (See Fig. 20.1.)

21.1 The Source of the Universe Problem

After considering the details of the calculation it appears that the fundamental problem was in the assumption that simplistic assumptions resulting in scale factors of the form of eqs. 11.2 and 11.3 were wrong. We find the Dark Energy term Ω_{Trest} (set to zero) is very large (as seen later) and the pressure is also very large and negative. *Dark Energy density (Ω_{Trest}) dwarfs the other densities listed in eq. 9.6 above. Thus the problems of the larger times model presented earlier. The Quantum Big Bang model still is valid. (see Fig. 21.1 below.)*

The missing ingredient in the use of the Einstein equation is the dynamic changes in the energy density $\rho_{tot}(t)$. This blank area has apparently been noted previously and Dark Energy has been suggested. Calculations of the amount of the Dark energy based on models such as quintessence have appeared. A noteworthy issue with many of these calculations that the suggest the Dark Energy density is a factor of 120 orders of magnitude lower than quantum field theory estimates.

In this chapter we will propose a universal scale factor that will lead to a Dark Energy that accounts for known features of the Hubble Constant as described in the Introduction to this book. (See Figs. 1 – 5.) It also implies extremely large Dark Energy density values that are not inconsistent with quantum field theory vacuum calculations. We will then see that the universal scale factor is the first phase of a universe evolution model that describes its overall evolution from the Big Bang metastate, with which it dovetails. Thus we have a complete phenomenology of the universe's evolution. The second phase of this model's theory (its justification – See section 4 of this book) is a calculation in quantum field theory for the case of a universe scalar particle with an electromagnetic-like interaction. This calculation leads to a vacuum polarization phenomena (as in our previous calculation of the QED fine structure constant.) It generates an eigenvalue function with a coupling constant eigenvalue that causes the universe particle to behave as a free field. The location of the eigenvalue specifies the power of the universal scale factor at short distances (early times after Fourier transformation).

Thus we obtain a theory of free universe particles (for our universe and other universes) with a specification of the evolution of the universe through a vacuum polarization-like process.

Time	0	t_c
Phase	**Big Bang metastate Beginning**	**Big Bang Metastate End**
Time	0	1.26×10^{-165}
Re a(t)	8.16×10^{-93}	1.632×10^{-92}
Im a(t)	-3.16×10^{-93}	-5.24×10^{-93}
Re radius	4.278×10^{-65} cm	8.5×10^{-65} cm
Im radius	4.278×10^{-65} cm	8.5×10^{-65} cm
Volume	4.37×10^{-192} cm^3	2.6398×10^{-191} cm^3
Central Expansion Energy Density	1.63×10^{218} GeV/cm^3	4.08×10^{217} GeV/cm^3
Edge Expansion Energy Density	8.16×10^{217} GeV/cm^3	2.04×10^{217} GeV/cm^3
Total Expansion Energy[97]	5.34×10^{35} eV	8.07×10^{35} eV
Hubble Constant[98] (km s^{-1} Mpc^{-1})	1.79×10^{218}	1.14×10^{126}

Figure 21.1 The Big Bang Metastate detailed data.

---

[97] The expansion energy does not include the mass-energy of particles in the Big Bang universe. It only includes Y field black body energy which drives the initial expansion. All particles are massless initially and all interparticle forces are zero.

[98] In the Big Bang metastable state we display the Hubble constant at the "expanding' edge.

22. The Universal Scale Factor for the Evolution of the Universe

In this chapter we define a universal scale factor that fits the known values of the Hubble Constant, that yields a value for the Hubble Constant consistent with our Big Bang metastate Hubble Constant values, and that has a form that makes physical sense.

The Robertson-Walker metric is

$$d\tau^2 = dt^2 - a^2(t)k^{-1}[dr^2/(1 - kr^2) + r^2(d\theta^2 + \sin^2\theta \, d\varphi^2)] \qquad (22.0)$$

a(t) is a function of time called the scale factor. Since a(t) is dimensionless there must be at least a second parameter with the dimension of time. We will call the parameter T. Thus the scale factor a(t) has the parameters t and T:

$$a(t, T)$$

although we will simply express it as a(t) generally.

This chapter calculates the growth of the universe based on a phenomenological approximation[99] to the growth of the universe from the Big Bang metastate to the present. It addresses the current impasse in our understanding of universe expansion.

Clearly something is wrong with our recent understanding of the universe. Experimental data tells us this.[100] A further suggestion of this problem appears in the negative values of the Hubble Constant squared in our calculations in chapter 20 of Blaha (2019c) which appears to be more than a result of approximations.[101]

---

[99] See Blaha (2019c). We establish a theoretic proof of the new scale factor in chapter 25 in section 4.
[100] See A. Riess *et al*, The Astrophysical Journal **875**, 145 (2019) and references therein.
[101] The primary cause of the problem seems to be an incorrect choice for energy density and pressure.

We begin with a look at the Hubble Constant expansion experimental data. We then proceed to suggest a phenomenology that accounts for the data and nicely projects back to the Big Bang metastate discussed earlier in Blaha (2019c) (and in Blaha (2004)). A justification for this phenomenological fit is provided in the following chapter 25 of section 4.

22.1 Hubble Constant Experimental Data

There are a number of astrophysical studies of the universe that suggest that the Hubble Constant is *not* constant. Although there are significant margins of error it appears that the early universe "beginning" epoch around 380,000 years had a Hubble Constant of 67.8 km s^{-1} Mpc^{-1}.[102] More recently, red shift studies of quasars have given a Hubble Constant of 73.2 km s^{-1} Mpc^{-1}.[103] And studies of binary black hole merger gravity waves[104] have given a Hubble Constant of 75.2 km s^{-1} Mpc^{-1} (and earlier of 78 km s^{-1} Mpc^{-1}). Another study of events at 1.8 billion ly yielded a Hubble Constant of 70.0 km s^{-1} Mpc^{-1}.[105] Further studies have given the Hubble Constants: 1) Of variable stars 73.2 km s^{-1} Mpc^{-1}, 2) Of light bent by distant galaxies 72.5 km s^{-1} Mpc^{-1}, 3) Of Magellan Cepheids 74.03 ± 1.42 km s^{-1} Mpc^{-1}, [106] 4) Of distant red giant[107] brightness 69.8 km s^{-1} Mpc^{-1},

The only apparent conclusion at this time is that there was a Hubble Constant (Constant) H of approximately 67.8 km s^{-1} Mpc^{-1} early in the universe, and ranging up to 75.2 km s^{-1} Mpc^{-1} at the current time. Thus an increasing Hubble Constant.

For the purpose of discussing the apparent increase in H with time, we average the above eight "recent" values of H in the spirit of Bayesian equal probability to obtain

---

[102] See, for example, K. Aylor *et al*, arXiv:1811.00537v1 (2018) based on studies of the cosmological sound horizon.
[103] M. Soares-Santos *et al* , arXiv:1901.01540 (2019).
[104] DES and LIGO collaborations *et al*, arXiv:1901.01540 (2019).
[105] B.P. Abbott *et al*, arXiv:1710.05835 (2017).
[106] J. T. Nielsen *et al*, Marginal evidence for cosmic acceleration from Type Ia supernovae, Nature Scientific Reports (2016); arXiv:1506.01354 (2015). A. Riess *et al*, The Astrophysical Journal **875**, 145 (2019) and references therein. A. Riess *et al*, arXiv:1903.07603 (2019).
[107] W. Freedman *et al*, The Astrophysical Journal **880** (July, 2019).

a **recent time Hubble average of 73.24**.[108] Thus there appears to be a 7% - 9% increase in the Hubble Constant over time.

22.2 Approaches to Universe Expansion

22.2.1 The Straightforward Approach

The standard approach to determining the evolution/expansion of the universe is to solve the Einstein equation

$$\dot{a}^2 - 8\pi G\rho_{tot}a^2/3 = -k \qquad (10.5.2)$$

using the energy density

$$\rho_{tot}(t) \equiv \rho_{crit}\Omega_{tot}(t) = \rho_{crit}[\Omega_\Gamma(t) + \Omega_M(t) + \Omega_\Lambda + \Omega_{TST} + \Omega_{Trest}]$$

$$= \rho_{crit}[\Omega_\gamma(t) + \Omega_\Lambda + \Omega_m(t) + \Omega_d(t) + \Omega_{Trest} + C_T G\rho_{Mega}\int_{t_T}^{t} dt'\rho_{tot}(t')/R(t')]$$

$$(9.6)$$

where Ω_{TST} is a surface tension estimate based Megaverse energy influx, and Ω_{Trest} is the additional Dark Energy which is set to zero. We did this in Blaha (2019c) and chapters 17 – 20. We found physically unsatisfactory results.

The cause, that we will see later, appears to be Dark Energy, Ω_{Trest}, that dwarfs the other terms in $\rho_{tot}(t)$. In the absence of an accurate estimate of Dark Energy density the "standard" approach cannot succeed.

22.2.2 Approach Based on a Quantum Vacuum Universe

In this chapter, and subsequent chapters, we will follow a different procedure: we will use the known features of the Hubble Constant to determine the scale factor a(t), and then use the Einstein equation to determine Ω_{Trest} and other quantities of

---

[108] In Blaha (2019c) and (2019d) we used an average estimate of 73.7.

interest. *Thus we will obtain estimates of the time varying Ω_{Trest}. It will prove to be enormous*[109] *and will lead us to consider Dark Energy.*

Thus the problem of Dark Energy is quantified by the use of the Einstein equation. The Einstein equation then becomes a consistency condition.

The second phase of this procedure is to justify the choice of the scale factor by a consideration of quantum vacuum polarization. This topic will be considered in chapter 25 in section 4 later. Our view is that the universe's expansion is guided by vacuum polarization.

22.3 A Phenomenological Fit to the Known Hubble Constant Data

The radiation-dominated and the matter-dominated scale factors a(t) both are power laws in time as seen earlier in Blaha (2019c). We therefore will assume that the true a(t) has a power law form:

$$a(t) = (t/t_{now})^{g + ht} \tag{22.1}$$

where g and h are constants. (The constant h is *not* the Hubble parameter.) There is an "ht" term in the exponent based on the rise in H(t) noted above in the experimental data.

The bases of this choice are:

1. Power law behavior (in part) as in the radiation and matter dominated approximations seen earlier.
2. The known shape of H(t) at early times, and at present, as seen and discussed in the Introduction. Eq. 22.1 is consistent with the shape of H(t).
3. The simplicity of the fit. Two values of H(t) set the constants g and h.
4. Faster than exponential future growth with no Big Rip.

$$a(t) = \exp[(g + ht)\ln(t/t_{now})] \sim e^{ht \ln(t)}$$

---

[109] Suggestions of an enormous Dark Energy have appeared earlier. Estimates tend to be 120 orders of magnitude below "back of the envelope" quantum vacuum theory estimates. We shall see later that the Dark Energy values that we calculate are many orders of magnitude more (for most times) than quantum vacuum theory estimates.

The Hubble Constant implied by eq. 22.1 is

$$H(t) = (da/dt)/a = g/t + h(1 + \ln(t/t_{now})) \qquad (22.2)$$

If we set the value of H(t) at two values of time, then g and h are determined. Based on the above discussion we use the experimental data:

$$H(t_c) \equiv H(380,000 \text{ yr}) = 67.8 \qquad (22.3)$$
$$H(t_{now}) = 73.24$$

Eqs. 22.2 and 22.3 imply

$$h = (t_c H(t_c) - t_{now} H(t_{now}))[\, t_c - t_{now} + t_c \ln((t_c/t_{now})]^{-1} \qquad (22.5)$$
$$g = (H(t_{now}) - h)\, t_{now}$$

Substituting the parameter values of eq. 22.3 we obtain

$$h = 2.25983 \times 10^{-18} \qquad (22.6)$$

$$g = 0.000282377 = 2.82377 \times 10^{-4}$$

Since

$$h \cong 1/t_{now}$$

an alternate possible form for a(t) is

$$a(t) = (t/t_{now})^{g + t/t_{now}} \qquad (22.1a)$$

We found this form to be suggestive but not consistent with the current values of the Hubble Constant. **We will use eqs. 22.1, 22.5 and 22.6 for calculations and in plots.**

    If one wanted the value of h to be equal to $1//t_{now}$ then the present value of the Hubble Constant would have to be 74.47(if the Hubble value at 380,000 years is 67.8) – a value within the range of experimental values.

We use these parameters to calculate a(t), and H(t) in the following plots in this chapter for the time ranges: 1) t = 0 to t = 10^{-165} s; 2) from the Big Bang metastate at t = 1.1984×10^{-165} s to the present; and 3) from 380,000 years to the present.

The times are displayed variously in seconds, $\log_{10}$ seconds, and in gigayears (Gyr). $\log_{10}$ plots were often necessary because of the wide ranges of values. *We now turn to comments on the plots since some new and hitherto unforeseen features emerge.*

22.4 Smooth H(t) Connection to Big Bang Metastate

The parameters in eqs. 22.1 and 22.2 were set by the H(t) data (eq. 22.5) at 380,000 years and the present. If we extrapolate back to the end of the Big Bang metastate then we find a good match between the values of a(t) and H(t) of the Big Bang metastate and the extrapolation:

	H(t)	
Big Bang Metastate	Big Bang Center	8.95×10^{217} km s^{-1} Mpc^{-1}
	Big Bang Edge	1.14×10^{126} km s^{-1} Mpc^{-1}

H(t) for the universal scale factor 9.149×10^{215} km s^{-1} Mpc^{-1} at t = 10^{-200} sec.
(and much larger as t → 0)

where the Big Bang values appear in section 13.6.2 of Blaha (2019c).. Note that the extrapolated value is within the range of values in the Big Bang metastate and is close (within a factor of 100) to the Big Bang Center value. Thus our H(t) fit (eqs. 22.1 and 22.2) extends smoothly back to the Big Bang. H(t) is in a rather remarkable approximate agreement for the Big Bang metastate.

22.5 Universal Scale Factor a(t) and H(t) Plots

The scale factor also shows a rough agreement. Using

$$\text{Re } a_{BBRW}(t_c) \cong \gamma = a(t_c) \qquad\qquad (13.3.2.17c)$$

from Blaha (2019c) where

$$\gamma = 1.632 \times 10^{-92}$$
$$t_c = 1.1984 \times 10^{-165} \text{ s}$$

we can see the removal of the singularity at t = 0 by combining the Big Bang metastate scale factor $a_{BBRW}(t)$ with a(t) in eq. 22.1.

For $0 \leq t \leq t_c$
$$a_{combined}(t) = \gamma/2 + a(t)/2 = \text{Re } a_{BBRW}(t) \tag{22.7}$$

For $t_c \leq t \leq t_{now}$
$$a_{combined}(t) = a(t) \tag{22.8}$$

Note a(0) = 0 while $a_{combined}(0) = \gamma/2 = 8.16 \times 10^{-93}$. Thus a Big Bang catastrophe is averted.

Fig. 22.1 plots $a_{combined}(t) = a(t)$, showing the Big Dip effect) from t = 10^{-200} sec. 1 sec. Fig. 22.2 plots $a_{combined}(t) = a(t)$ from t = 1.19×10^{13} sec. (380,000 years) to the present (4.35 × 10^{17} sec). Fig. 22.3 $a_{combined}(t) = a(t)$ from t = 1.19×10^{13} sec. (380,000 years) to the present (4.35 × 10^{18} sec.) projecting into the future.

The plots of H(t) show the Big Dip, the past and the future:

Fig. 2.4: Log_{10} H(t) vs. log_{10} t sec. from t = 10^{-200} sec. to t = 1 sec.

Fig. 2.5: Log_{10} H(t) vs. log_{10} t sec. from t = 1.19×10^{-164} sec. to t = 4.31×10^{67} sec. – The distant future. No Big Rip!

Fig. 2.6: H(t) vs. t sec. from t = 1.19×10^{13} sec. (380,000 years) to t = 1.31×10^{18} sec. showing the Big Dip and a rise into the future (roughly three times the current age of the universe).

Fig. 2.7: H(t) vs. t sec. from t = 1.19×10^{13} sec. (380,000 years) to t = 4.35×10^{17} sec. (the present) showing the Big Dip. A closeup of the Big Dip region.

22.6 Fluctuating Behavior of a(t) and H(t) – The Big Dip

An examination of the below figures reveal a Big Dip in H(t) which seems to have been unforeseen in astrophysical investigations. The Big Dip has a corresponding big peak in the total additional energy density $\Omega_T(t)$ (chapter 23.):

$$\Omega_T(t) = \Omega_{TST} + \Omega_{Trest} \qquad (22.9)$$

The big peak occurs at approximately the same point in time as the Big Dip in H. There is a corresponding minimum for a(t).

The cause of the Big Dip is the form of the universal scale factor. It would be present (although slightly modified) even if the Hubble Constant were truly constant from the 380,000 year point to the present.

The reason for these anomalies is somewhat uncertain although it is apparent that they are not a fault of the universal scale factor in eq. 22.1.

22.6.1 Location of the Big Dip in H(t)

The locations in time of the Big Dip events are:

> Big Dip low point (H = -445) at t = 8.71×10^{13} sec.
> Ω_T plot peak at t = 8.71×10^{13} sec   (See chapter 23)
>
> Big Dip in a(t) has the value 0.69628 at t = 1.56×10^{17} sec.

The decrease in a(t) is to 69.6% of the initial value of 1. See Fig. 22.3. The delay from the H(t) Big Dip value to the low point of a(t) can be attributed to the time required to propagate the diminished H value throughout the universe.

Since the H and Ω_T times match it appears that the Big Dip is a result of massive growth in the mass-energy (and pressure) embodied in Ω_T. See chapter 23.

Since the changeover from a radiation-dominated phase to a matter-dominated phase occurs at a slightly earlier time:

Radiation – Matter Domination Transition:[110] $t = 1.48 \times 10^{12}$ sec.

it seems reasonable to conclude the transition from radiation-dominated to matter-dominated causes the Big Dip to occur. The matter-dominated phase transition causes shrinkage as shown in a(t) in Fig. 22.3. *The universe contracts by one-third!* [111] We attribute the time delay between the transition and the Big Dip in a(t) to the time required for the transition to occur. (The universe is large at this time after all)

22.7 Universe Contraction – Early Massive Galaxies

The contraction would appear to "squeeze" the mass-energy in the universe giving it a "belly" (the Big Belly??? of "squeezed" mass-energy).This mass-energy contraction leads to the early formation of galaxies that disperse due to gravitation in the 13.5 Gyrs that follow. The subsequent expansion would also appear to create a "wake" similar to the wake of a water wave.

Evidence[112] has been found for the existence of a huge population of very massive galaxies (39+ have been found so far) that were created within one billion years after the Big Bang. This population of early galaxies is inconsistent with the standard present-day models of galaxy formation. The Big Dip occurs at 2.76 million years – well before one billion years – consistent with the formation of early massive galaxies.

A concentration of mass-energy due to the contraction of the universe appears to present a possible solution. Universe contraction was not considered in the creation of these models of galaxy formation.

Another possible source of universe concentrations (and voids) of energy appears in our Quantum Big Bang Model seen earlier. The cause is a large difference in expansion rates (Hubble Constant variations) at the center of the Big Bang compared to the outer edge of the Big Bang as shown earlier in the section 2 Big Bang Model.

---

[110] In view of our universal scale factor formulation the time of the radiation-matter transition becomes questionable.
[111] Rather like the condensation of water vapor to liquid.
[112] T. Wang *et al*, Nature **572**, 211 (2019).

22.7.1 Overshoot in H(t)

The result of the Radiation-Matter transition seems to be a negative H(t) for the energy density Ω_T. H(t) "overshoots" and becomes negative. Crudely put, the clumping of matter in the matter-dominated phase appears to introduce a compactness that results in a decrease in universe size and concentrations of energy.

22.8 Voids and Bubbles in Space after the Big Dip

The Big Dip concentrates mass-energy at the contraction. The subsequent expansion creates a "type" of wave that generates massive galaxies (bubbles of mass-energy), and also voids – bubbles of space devoid of galaxies. In the course of the following thirteen or so billion years gravitation causes a dispersion of galaxies, voids and bubbles leading to the present day observed distribution.

22.9 Mystery of the Big Dip in H(t) - A Scenario

At the Big Dip H(t) changes from a declining to a rising trajectory. Based on this fact and the Big_Bang-Megaverse_Driven model presented in Blaha (2019c) the following scenario seems reasonable:

1. The initial peak, and immediate decline, in H(t) is due to the Y black body radiation phase pressure that decreases rapidly after the Big Bang metastate ends. Thus Ω_T declines rapidly (See chapter 23.) with the Y pressure decline. (Note Ω_T is a sum of energy density and pressure.)

2. The peak in Ω_T reflects an influx of energy (from the Megaverse?) that causes H(t) to begin increasing. There is also a dip below zero in H(t) in Figs. 22.6 and 22.7 signifying the shrinkage of the universe as a(t) in Fig. 22.2 shows.

3. Afterwards Ω_T continues to be significant and increasing as a(t) and H(t) rise to the present time. See chapter 23.

4. In the future Ω_T should continue to rise. The energy increase that this situation implies suggests a certain reality to the part of our Big_Bang-Quantum_Vacuum Theory. See Section 4 for details.

22.10 Current H(t() Increase Explained

The scenario presented in the previous section "explains" the rise of H(t) from 67.8 to an average 73.24.

22.11 An Interlude in the Eons

Based on the above analysis it appears that the universe is currently in an *interlude* following a decline in growth rate after the Big Bang, and the new beginning of major growth due to Quantum Vacuum effects that may last "forever."

22.12 Scaling in the Scale Scale Factor: Changing the Base Time from t_{now} to a different Base Time

The use of $T = t_{now}$ raises a question of the dependence of a(t) and the other quantities on the current time. To remove that impression we will express the previous results in terms of a different base time T' instead of t_{now}.

$$a(t, T) = a'(t', T') = (t/T')^{g' + h't'} \qquad (22.10)$$

where

$$t' = tT/T' \qquad (22.11)$$
$$g' = g$$
$$h' = hT/T'$$

Although the parameterization appears different the physical results are the same.[113]

---

[113] In that regard it is analogous to the renormalization group analysis of coupling constants with "coupling constants" h =h(t_{now}), h' = h(t_m), and the times t_{now} and t_m being analogous to energy scales. See section 4.

22.13 Why is the Fit So Successful?

The universal scale factor in eq. 22.1 appears to be a successful fit to the growth of the universe. It raises the question: Why is it so successful? It matches the Big Bang model at the beginning. It accounts for the observed rise in H(t). It shows an initial decline in H(t) after the end of the Big Bang epoch. It matches our expectation that the rise in H(t) is the result of an energy inflow from the Megaverse.

In this author's view its success is probably attributable to renormalization group self-similarity.[114] The evolution of universes may have laws of growth from a Big Bang metastate that are based on self-similar growth along the lines of the Renormalization group. Chapter 25 of section 4 discusses a close analogy with quantitative similarities between electron vacuum polarization and the form of the growth of the universe.

Note: all logarithms in the following plots are to base 10.

---

[114] The self-similarity is based on the principle that the physics of the evolution of universes should be independent of the choice of the present time t_{now}. A change in t_{now} should result in a corresponding change in parameters—the essence of the Renormalization group.

a(t)

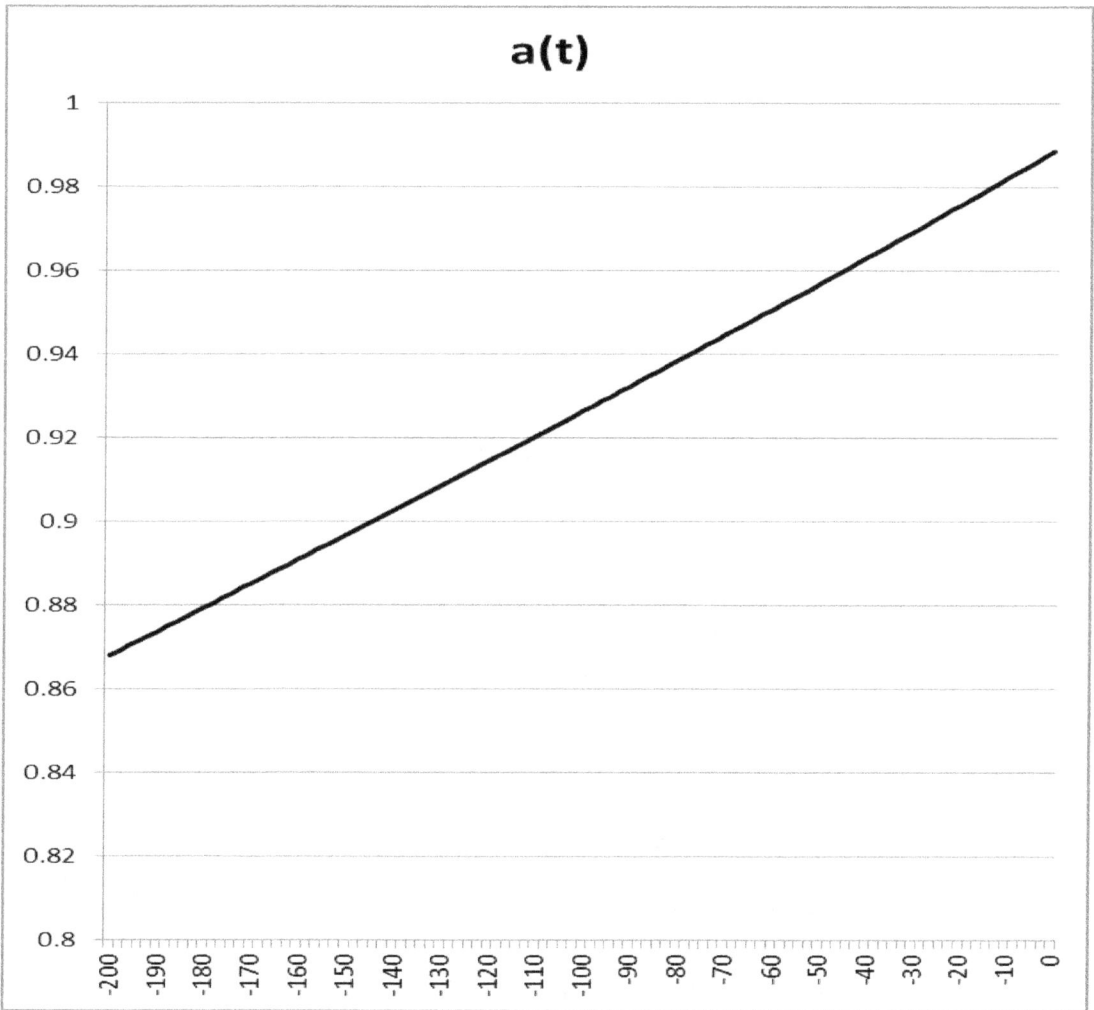

Figure 22.1. Universal Scale Factor a(t) plotted in $\log_{10}$ seconds from t = 10^{-200} sec. to t = 1 sec.

a(t)

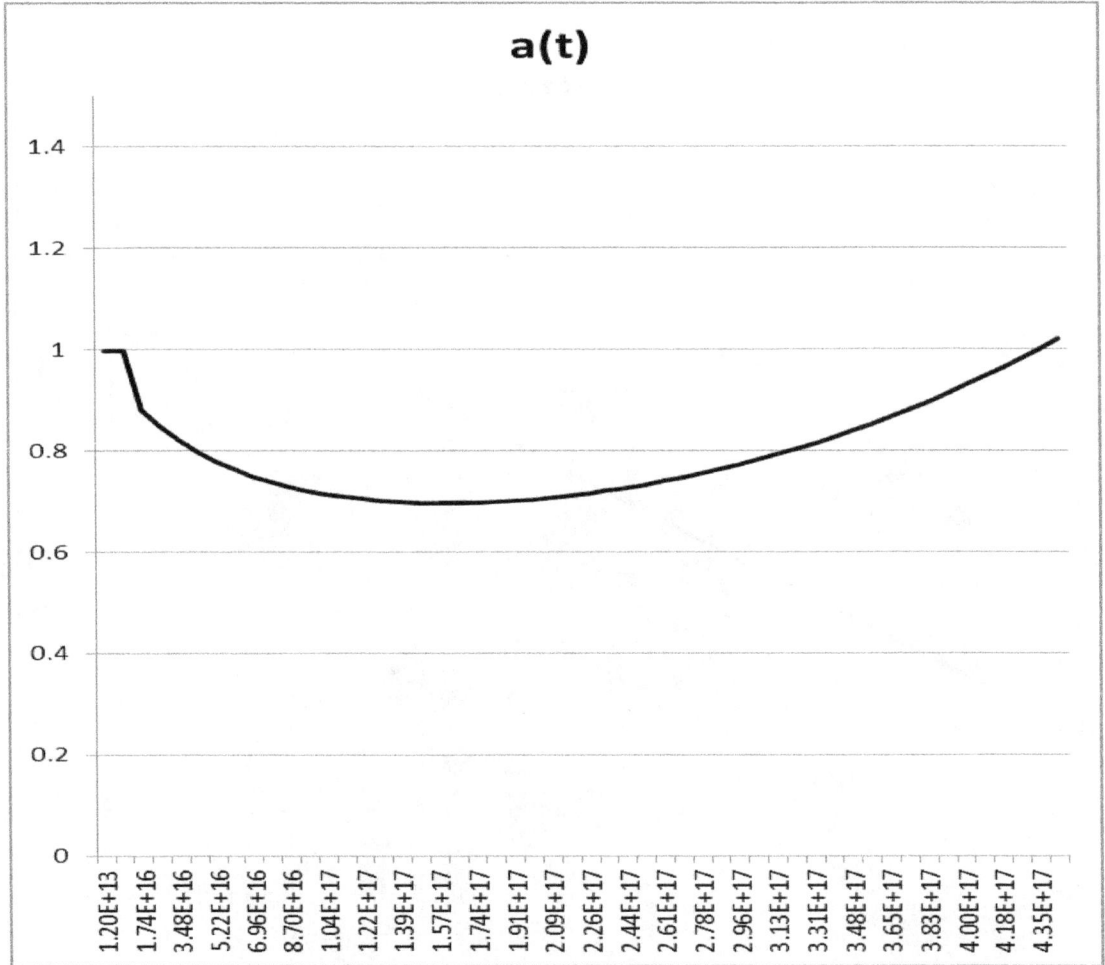

Figure 22.2. Universal Scale Factor a(t) plotted in seconds from t = 1.19×10^{13} sec. (380,000 years) to t = 4.35×10^{17} sec. (the present). Note the Big Dip to a = 0.69628 at t = 1.56×10^{17} sec. The decrease in a(t) is to 69.6% of the initial value of 1. (The almost flat part before 1.20×10^{13} sec is a roundoff effect. See Fig. 22.3.)

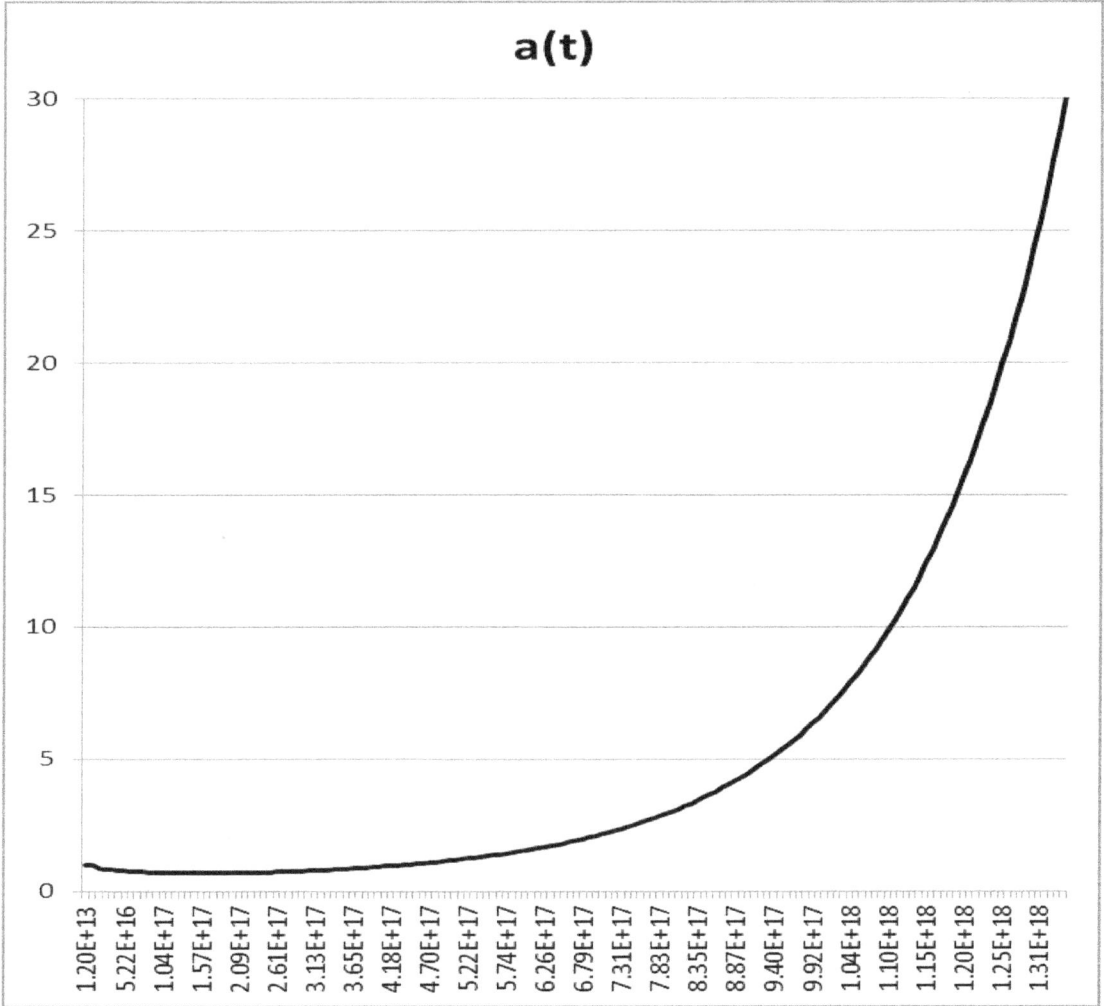

Figure 22.3. Universal Scale Factor a(t) plotted in seconds from t = 1.19×10^{13} sec. (380,000 years) to t = 1.31×10^{18} sec. (the distant future). Note the Big Dip appears diminished due to the scale of the plot.

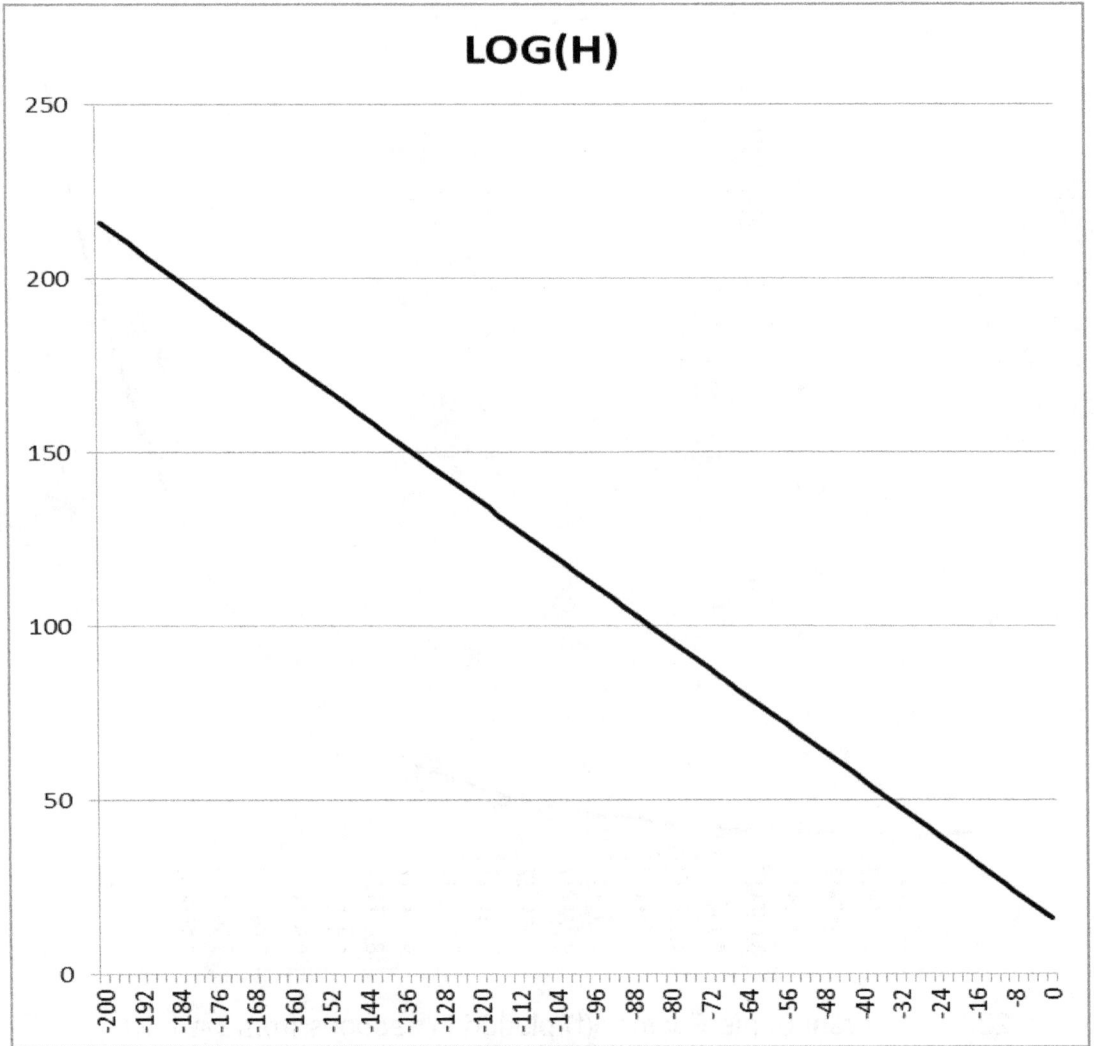

Figure 22.4. Log_{10} H(t) plotted in log_{10} seconds from t = 10^{-200} sec. to 1 sec.

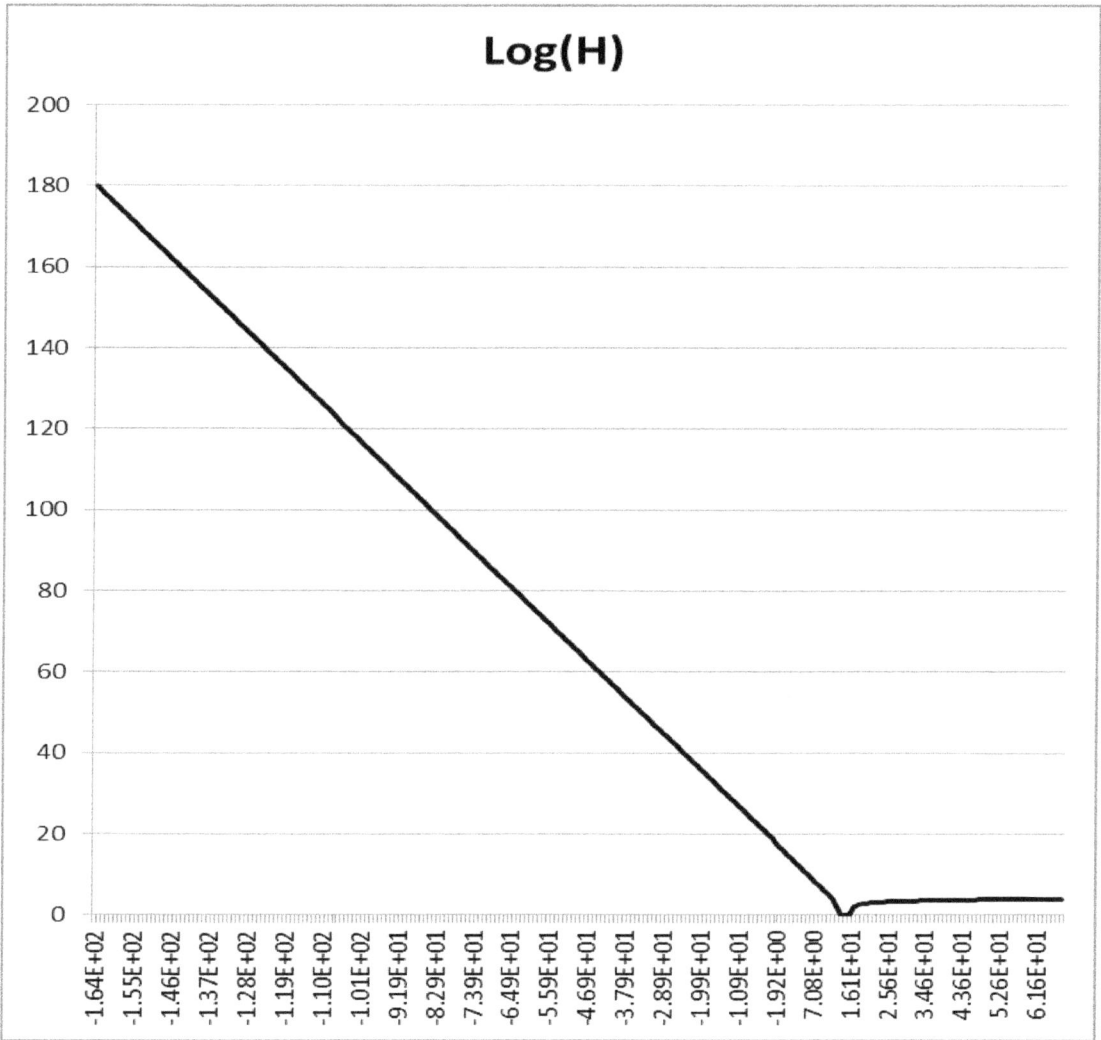

Figure 22.5. Log_{10} H(t) plotted in log_{10} seconds from t = 1.19×10^{-164} sec. to the distant future 4.31×10^{67} sec.

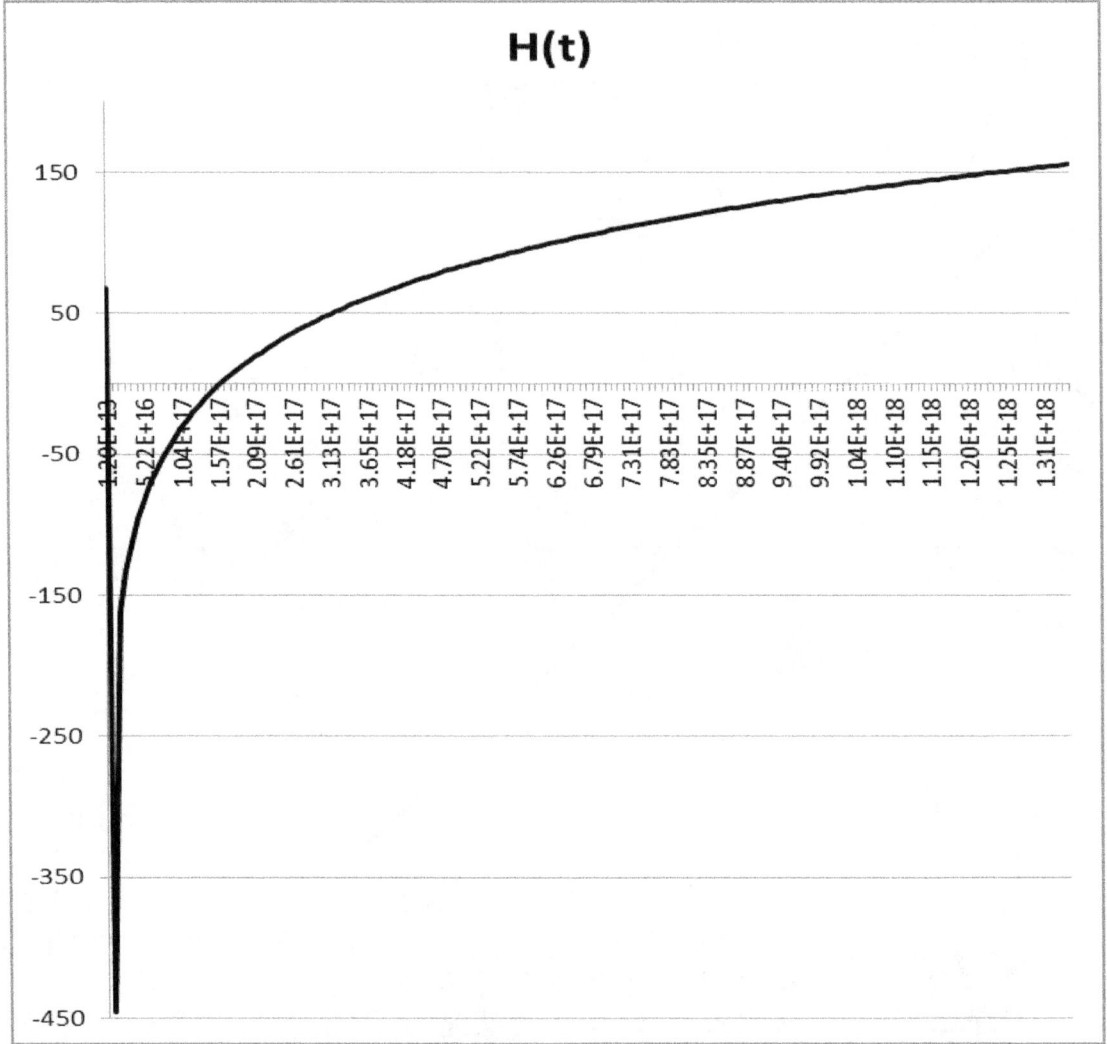

Figure 22.6. H(t) plotted vs. seconds from t = 1.19×10^{13} sec. (380,000 years) to 1.31×10^{18} sec. (three times the present time) showing the Big Dip in H(t).

H(t)

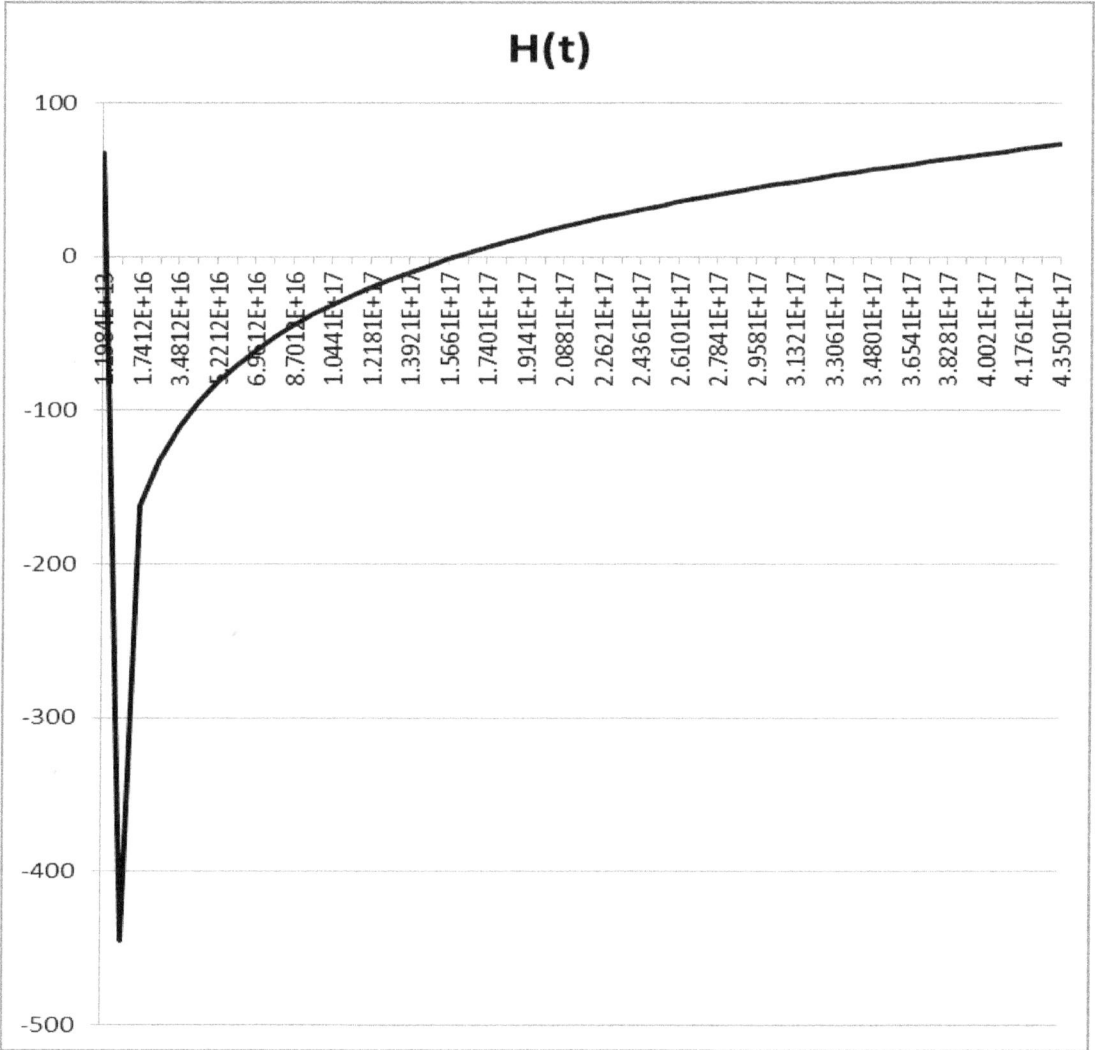

Figure 22.7. A closer view of H(t) plotted vs. seconds from t = 1.19 × 10^{13} sec. to the present 4.35 × 10^{17} sec. The minimum is H = -445 km s^{-1} Mpc^{-1} at t = 8.71 × 10^{13} sec.

23. Energy Density, Pressure, Dark Energy, and Equation of State Implied by the Universal Scale Factor

The universal scale factor defined in chapter 22 implies the time dependence of the universe's Energy Density, Pressure, Dark Energy, and Equation of State. In this chapter we calculate these quantities and then plot their values as a function of time.

We start with the expressions for the universal scale factor and its derivatives:

$$a(t) = (t/t_{now})^{g + ht} \tag{22.1}$$

$$da(t)/dt = a[g/t + h + h \ln(t/t_{now})] \tag{23.1}$$

and

$$d^2a(t)/dt^2 = da/dt \, [g/t + h + h \ln(t/t_{now})] + a(h - g/t)/t$$

23.1 Dark Energy

The Einstein equation

$$\dot{a}^2 - 8\pi G\rho_{tot}a^2/3 = -k + \Lambda \, a^2c^2/3 \tag{10.5.2}$$

enables us to determine the dark energy energy) $\Omega_T(t)$ beyond the cosmological constant, using the energy density:

$$\rho_{tot}(t) \equiv \rho_{crit}\Omega_{tot}(t) = \rho_{crit}[\Omega_\Gamma(t) + \Omega_M(t) + \Omega_\Lambda + \Omega_T(t)] \tag{23.2}$$

Thus

$$\Omega_T(t) = (H(t)^2 + c^2k/a^2)/H_0^2 - \Omega_\gamma/a^4 - \Omega_M/a^3 - \Omega_\Lambda \tag{23.3}$$

where

$$H_0^2 = 8\pi G\rho_{crit}/3$$

The quantity Ω_T is the excess of energy beyond that specified by Ω_γ, Ω_M, and Ω_Λ in the conventional expressions for the mass-energy density. The calculation of Ω_T which is in part energy density, (and possibly in part a Megaverse energy influx), is based on eq. 23.3 using the Einstein equation in the form

$$H(t) = (da/dt)/a(t) = [H_0^2 \rho_{tot}(t)/\rho_{cri} - c^2 k/a^2(t)]^{1/2}$$

Figs. 23.1, 23.2, 23.4 and 23.5 plot $\Omega_T(t)$ and its derivative. It is clearly much larger than $\Omega_H(t)$ (Fig. 23.2) showing why the standard approaches to calculating a(t) using $\Omega_H(t)$ are faulty. The ratio Ω_H/Ω_T is appreciable only after 380,000 years. See Fig. 23.14.

$$\Omega_H = \Omega_\gamma(t) + \Omega_\Lambda + \Omega_m(t) + \Omega_d(t) \qquad (21.2)$$
$$\Omega_H = \Omega_T(t) + \Omega_\Lambda + \Omega_M(t)$$

23.2 Energy Density

The energy density is given by eq. 23.2. It is plotted in Figs. 23.6, 23.7 and 23.8. Fig. 23.6 shows a drop below zero starting at t = 0.12 sec. ($\log_{10}$ t = -0.92) signaling very low values for ρ. There is a minimum at t = 1.2×10^{16} sec. in Fig. 23.7.

Note also the drop below zero in Fig. 23.6 beginning at t = 5×10^{-16} sec. signaling the start of very small values for ρ.

The derivative of total energy density $d\rho_{tot}/dt$ is plotted in Fig. 23.8.

23.3 Pressure

The Friedmann-Lemaître-Robertson-Walker metric equations yield the pressure

$$p = - (c^2 \rho_{crit}/(3H_0^2))[2(d^2a/dt^2)/a + (da/dt)^2/a^2 + kc^2/a^2 - \Omega_\Lambda] \qquad (23.4)$$

using eqs. 23.1 and 23.2 above.

An alternate approach to determining the pressure p uses the energy conservation equation

$$d(\rho_{tot}R^3)/dR = -3pR^2 \qquad (23.5)$$

where

$$R = k^{-\frac{1}{2}} a(t) \tag{23.6}$$

and gives

$$p = - (da/dt)^2 \rho_{tot} - a\, da/dt\, d\rho_{tot}/dt \tag{23.7}$$

$$p = -a(t)^2(H(t)^2\rho_{tot} + 1/3\, H(t)\, d\rho_{tot}/dt\,)/[(2.85 \times 10^{37})c^2] \tag{23.8}$$

where $k = 5.56 \times 10^{-57}$ cm^{-2}, ρ is the total energy-mass density, p is the pressure, and the denominator is required by the dimensions to obtain p in gm/cm^3. See Fig. 23.9, 23.10 and 23.11.

In Fig. 23.9 note the low values of $\log_{10}(-p(t)) \approx -115$ beginning at $\log_{10}(t$ sec.$) = 12.1$ ($t = 1.2 \times 10^{12}$ sec.). The pressure is always negative indicating a pressure for expansion. At $t = 1.19 \times 10^{-165}$ sec. we find $\log_{10}(-p(t)) = 593$ or $p(t) = -10^{593}$ gm/cm^3.

In Fig 23. 11 the maximum pressure in the range of 380,000 years to the present is -2.62×10^{-124} gm/cm^3 at $t = 1.0 \times 10^{17}$ sec.

23.4 Derivative of Dark Energy

The derivative of the Dark Energy, which we identify as $d\Omega_T/dt$ is

$$d\Omega_T/dt = H(t)[-2g/(H_0^2t^2) +2h/(H_0^2t) - 2c^2k/a^2 +4\Omega_\gamma/a^4 + 3\Omega_M/a^3] \tag{23.9}$$

We find that it is well-approximated by

$$d\Omega_T/dt \approx H(t)[-2g/(H_0^2t^2) +2h/(H_0^2t) - 2c^2k/a^2\,] \tag{23.10}$$

Note the dip below zero in Fig. 23.5 beginning at $t = 1.2 \times 10^{10}$ sec. – somewhat earlier than the Big Dip.

23.5 Dominance of Dark Energy

Similarly we find eq. 23.3 is well approximated by

$$\Omega_T(t) \approx (H(t)^2 + c^2k/a^2)/H_0^2 \tag{23.3'}$$

due to the relative smallness of Ω_H. Dark Energy dominates.

The total energy density is

$$\rho_{tot}(t) \equiv \rho_{crit}\Omega_{tot}(t) = \rho_{crit}[\Omega_\Gamma(t) + \Omega_M(t) + \Omega_\Lambda + \Omega_T(t)] \qquad (23.11)$$

See Fig. 23.6. It is well approximated by

$$\rho_{tot}(t) \approx \rho_{crit}\Omega_T(t) \qquad (23.12)$$

23.6 Equation of State

The equation of state

$$w = p/\rho$$

is plotted in Figs. 23.12 and 23.13 as a function of time. Fig. 23.12 shows an enormous range of values from the Big Bang to the present. Fig. 23.13 shows that w is of the order of 10^{-95} for most of the interval from 380,000 years to the present. At its "peak" Fig. 23.13 shows $w = -1.36 \times 10^{-95}$ at the time $t = 1. \times 10^{17}$ sec.

Since w is approximately zero in these time intervals it can be viewed as describing cold dust or gas.

Since quintessence is viewed as indicated by $w \neq -1$, the theory has quintessence.

23.7 Deceleration Parameter q

The deceleration parameter q is plotted in Figs. 23.15 and 23.16.

$$q = -ad^2a/dt^2/(da/dt)^2 \qquad (23.13)$$

It is proportional to the second derivative of the universal scale factor.

If $q < 0$ then it indicates accelerating expansion of the universe. If $q > 0$ then it indicates a decelerating universe.

Figs. 23.15 and 23.16 both have a pronounce maximum and minimum. The maximum occurs at $t = 1.2 \times 10^{13}$ sec. The minimum occurs at 1.6×10^{17} sec. To the left of the maximum (early times) $q > 0$ indicating decelerating expansion. To the right of the maximum for $t > 1.7 \times 10^{14}$ sec. we find $q < 0$ indicating accelerating expansion The accelerating expansion started 13.78 billion years ago – "just" after the Big bang. At present $q = -2.0$ – accelerating expansion – as suggested by astrophysical experiments. In the future, at $t = 8.2 \times 10^{17}$ sec. we found $q = -1.2$. The accelerating expansion will continue.

23.8 Comments on Universal Scale Factor Quantities

The complete universe "life" history presented in the following plots is consistent with a declining pressure, a declining density, and a declining pressure from the Big Bang phase consistent with our physical expectations. The Big Dip and subsequent rises would appear to be due to a sharp influx of energy (Fig. 23.2) possibly from the Megaverse.

In Fig. 24.1 of chapter 24 we present a time series of the evolution of the universe.

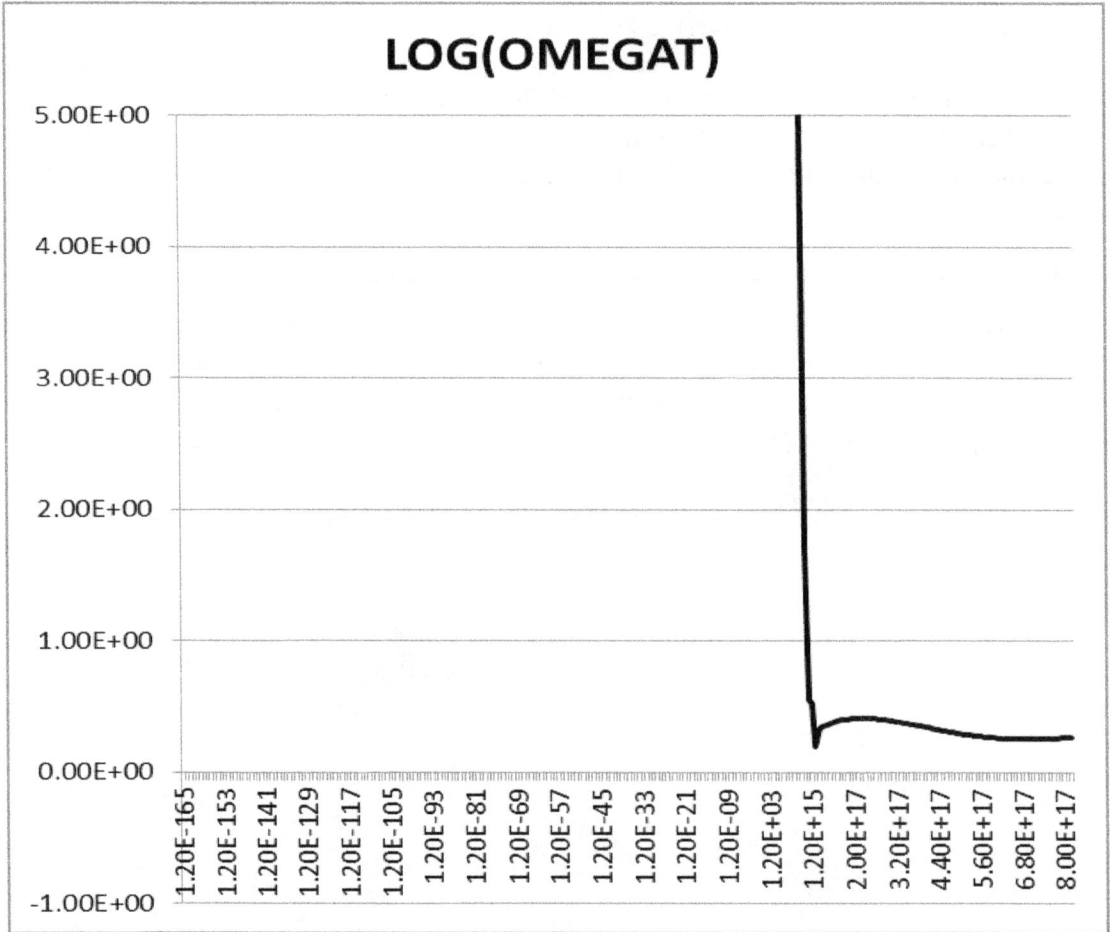

Figure 23.1. Log $\Omega_T(t)$ plotted vs. log time in seconds from 1.19×10^{-165} sec. to the 8.2×10^{17} sec. (almost double the present time). Note the Big Dip in $\Omega_T(t)$ at about $t = 8.71 \times 10^{13}$ sec. followed by a rise then a decline. At $t = 1.19 \times 10^{-165}$ sec. we found (not shown) log $\Omega_T(t) \approx 358$, an enormous value, corresponding to the level of vacuum energy found in quantum field theory. It declines to the plotted data shown above.

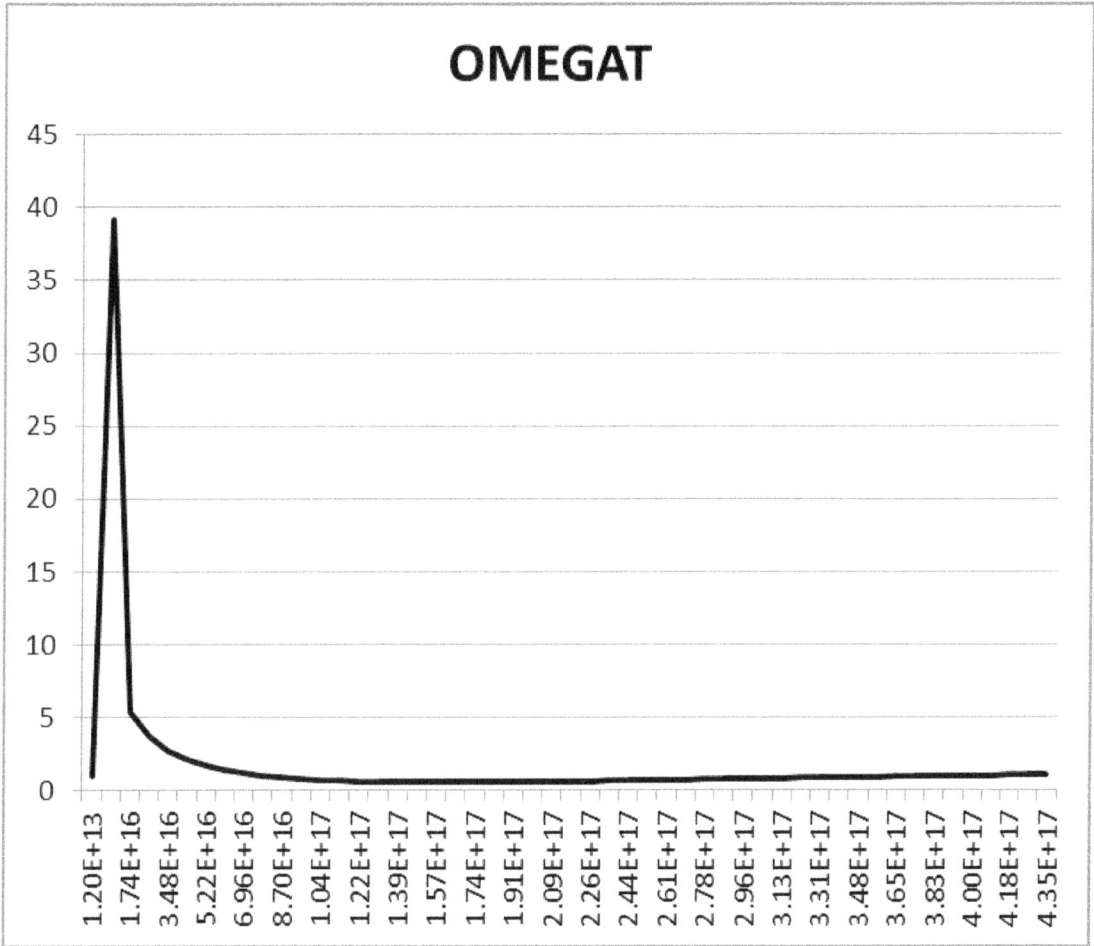

Figure 23.2. $\Omega_T(t)$ plotted vs. time in sec. from the year 380,000 to the present. Note the peak at t = 8.71×10^{13} sec. suggesting an influx into the universe at the transition to the matter-dominated phase, and then a decline followed by a raise. The peak value of $\Omega_T(t)$ is 39.2.

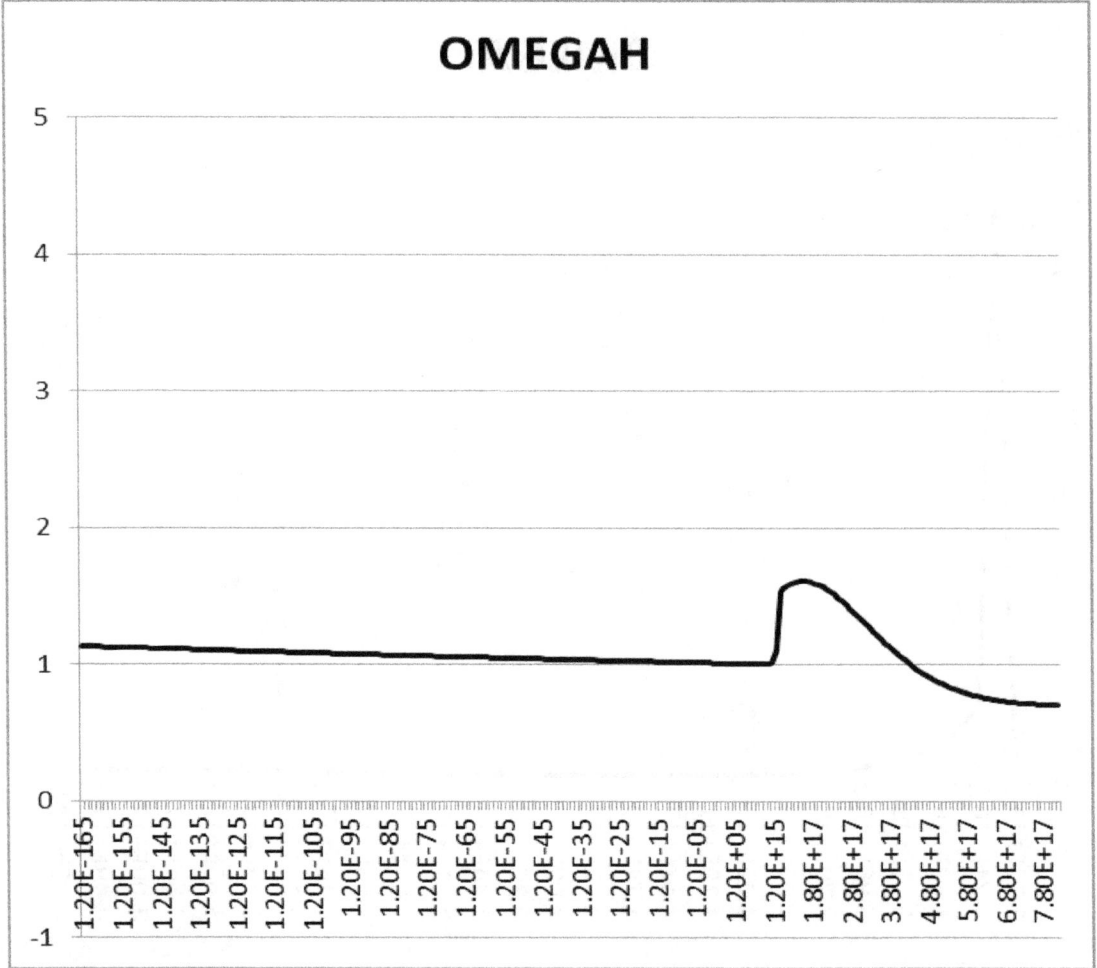

Figure 23.3. $\Omega_H(t)$ plotted vs. log time in seconds from 1.19×10^{-165} sec. to the 8.2×10^{17} sec. (almost double the present time). Note the start of a rise at t = 1.2×10^{13} sec. which coincides with the start of the appearance of atoms at t_T = 380,000 years = 1.19837×10^{13} sec.

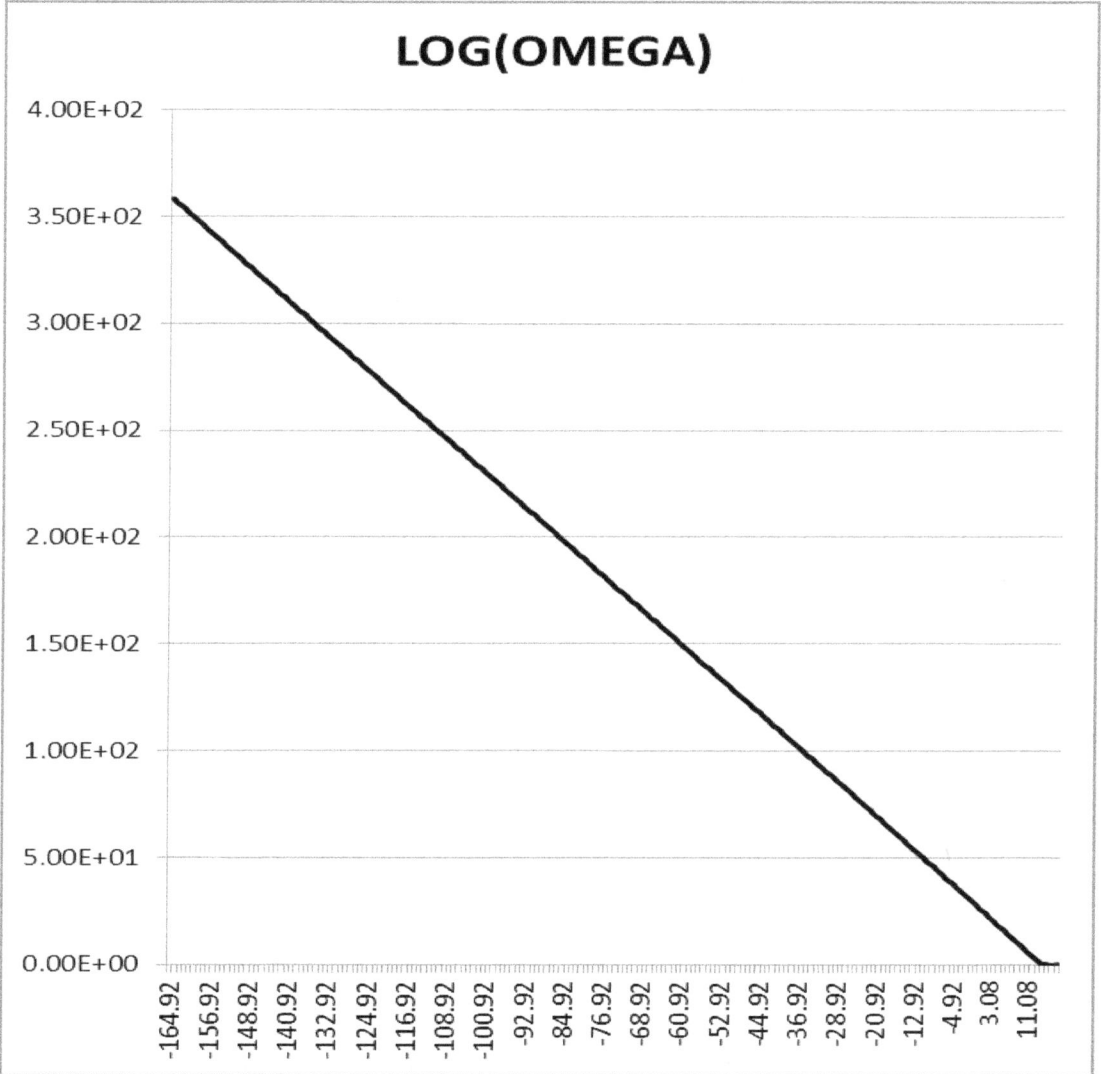

Figure 23.4. $\text{Log}_{10}\ \Omega_T(t)$ plotted vs. $\log_{10} t$ sec. from the Big Bang metastate to the present. Note the "dip" below zero of $\Omega_T(t)$ at $\log_{10} t = 11$.

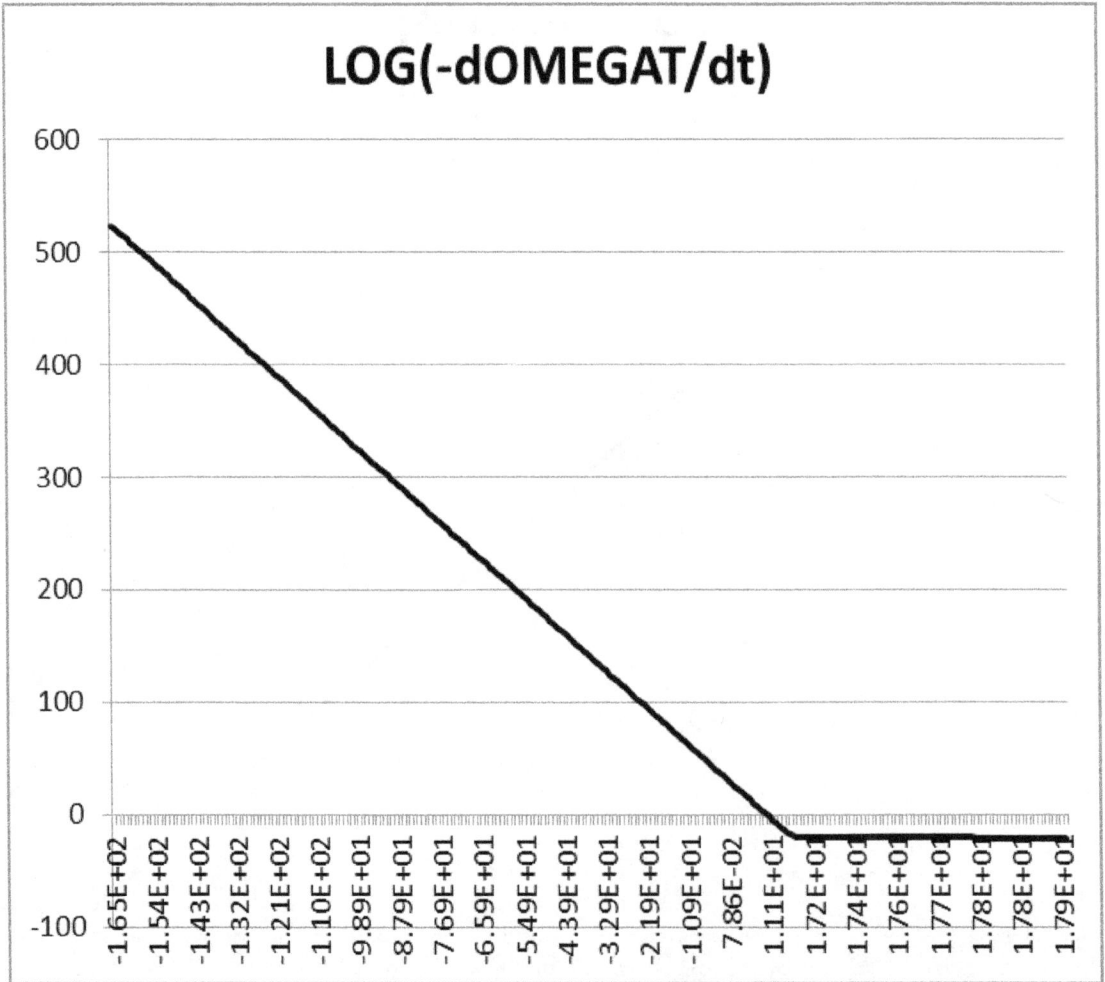

Figure 23.5. $\text{Log}_{10}(-d\Omega_T(t)/dt)$ plotted vs. $\log_{10}(t)$ sec. from the Big Bang metastate $t = 1.19 \times 10^{-165}$ sec. to the future: $t = 8.2 \times 10^{17}$ sec. Note dip below zero beginning at $t = 1.2 \times 10^{10}$ sec. – somewhat earlier than the Big Dip.

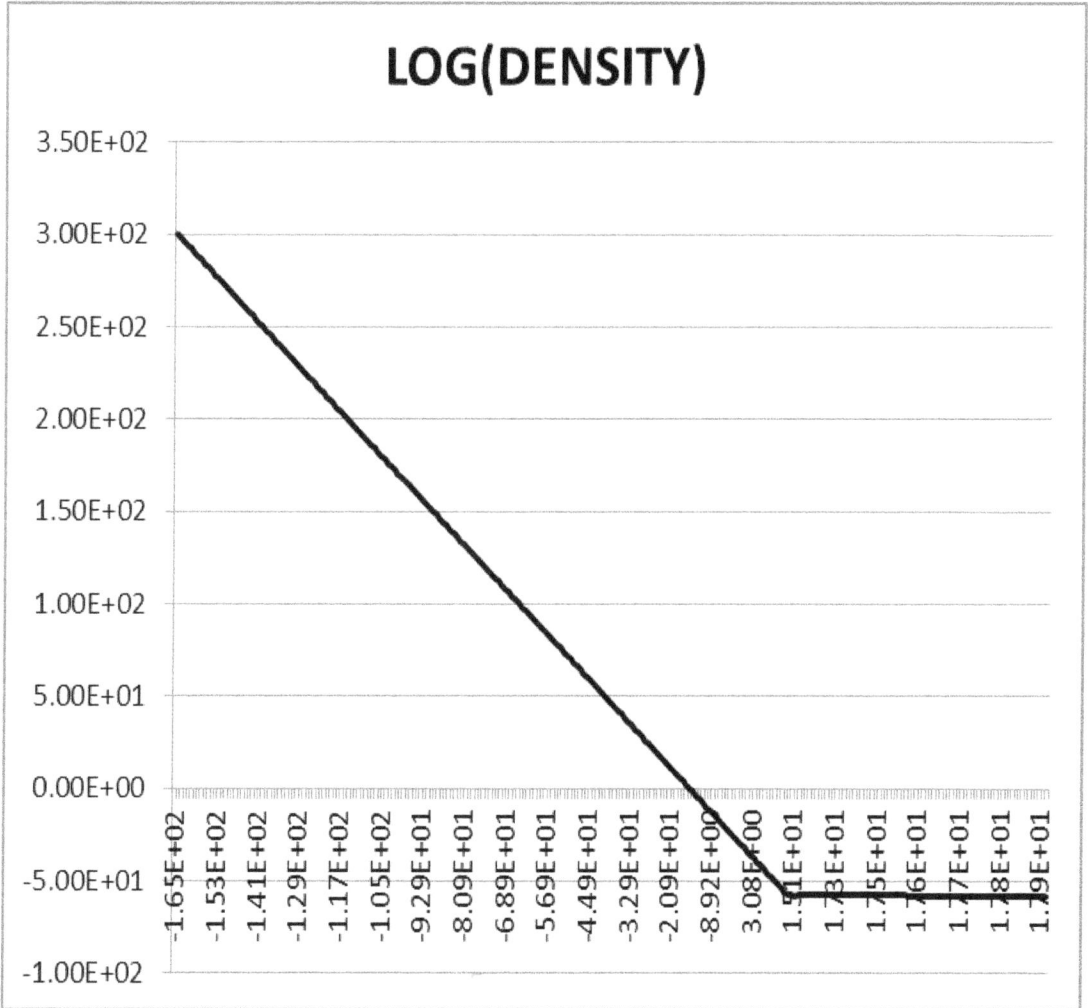

Figure 23.6. Log Density: $Log_{10}(\rho_{tot}(t)$ g/cm$^{-3})$ plotted vs. $log_{10}(t)$ from the Big Bang metastate at t = 1.19×10^{-165} sec. to the future: t = 8.2×10^{17} sec. Note the drop below zero beginning at t = 5×10^{-16} sec. signaling the start of very small values for ρ.

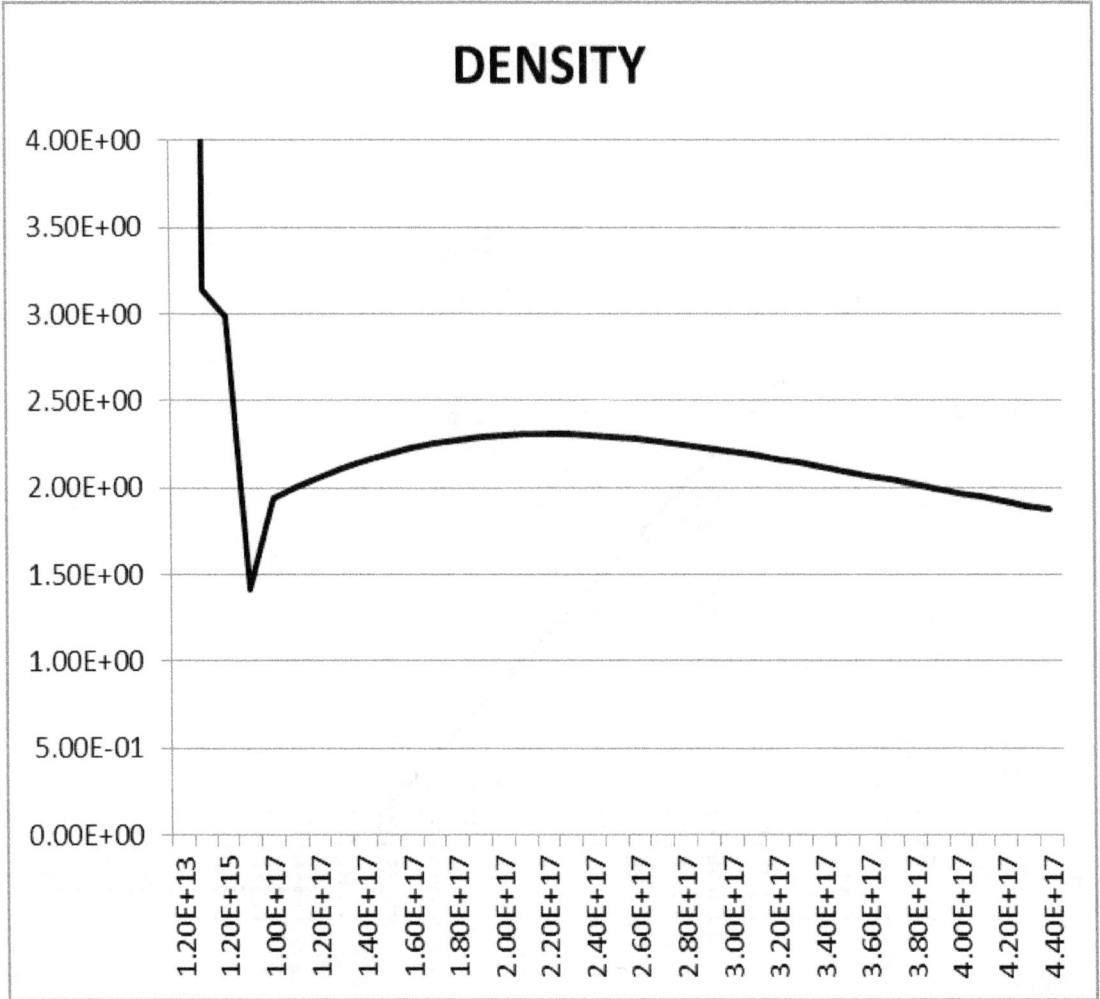

Figure 23.7. Energy density $\rho(t)$ g/cm^{-3} $\times$ 10^{29} plotted vs. t from t = 1.19 $\times$ 10^{13} sec. (380,000 years) to the present 4.35 $\times$ 10^{17} sec. Note the dip at t = 1.2 $\times$ 10^{16} sec. with a small value for ρ = 1.42 $\times$ 10^{-29} g/cm^{-3}.

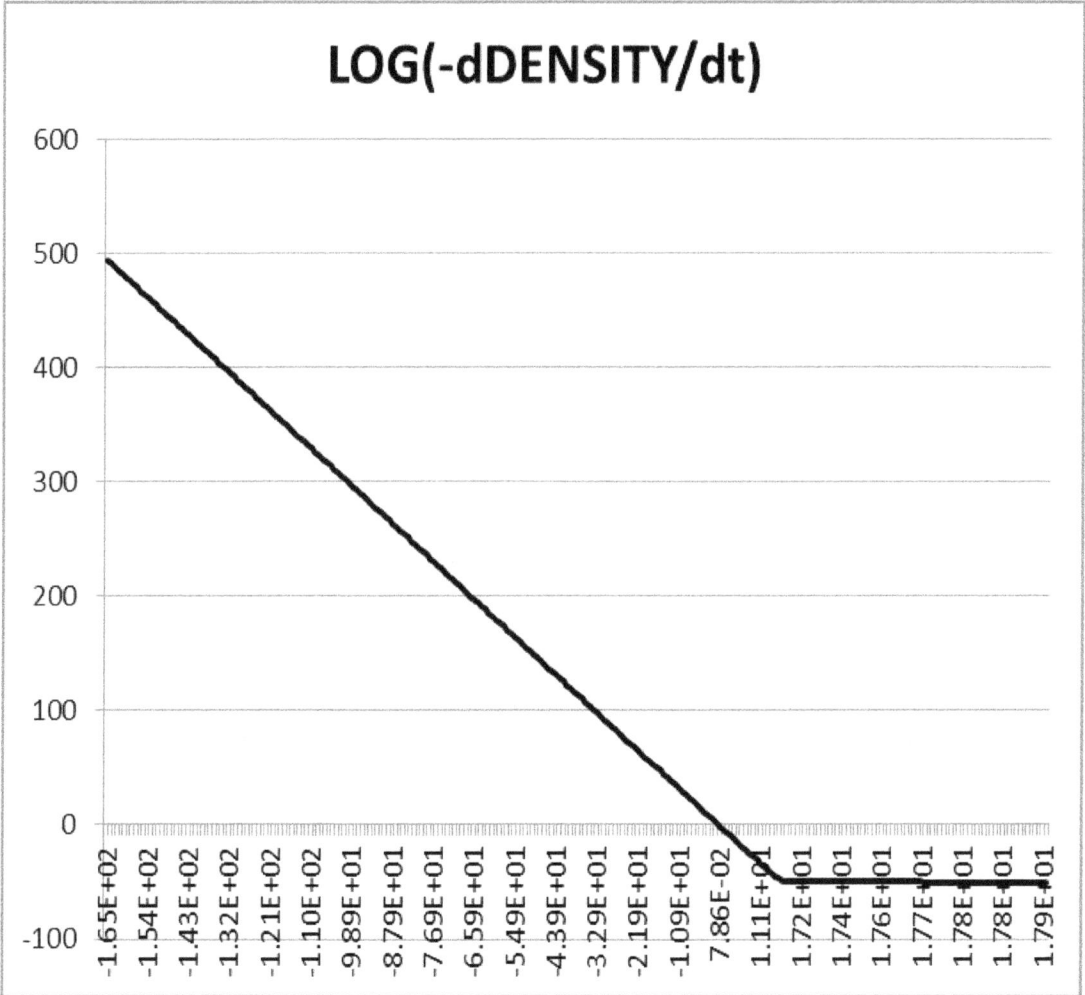

Figure 23.8. $\text{Log}_{10}(-d\rho(t)/dt \ g \ s^{-1} \ cm^{-3})$ plotted vs. $\text{log}_{10}(t)$ from the Big Bang metastate at $t = 1.19 \times 10^{-165}$ sec. to the future: $t = 8.2 \times 10^{17}$ sec. The plot goes below zero at about $\text{log}_{10}(t \ sec.) = 7.8 \times 10^{-2}$ ($t = 1.198$ sec.) where $\text{log}_{10}(-d\rho/dt \ g \ s^{-1} \ cm^{-3}) = -0.76$.

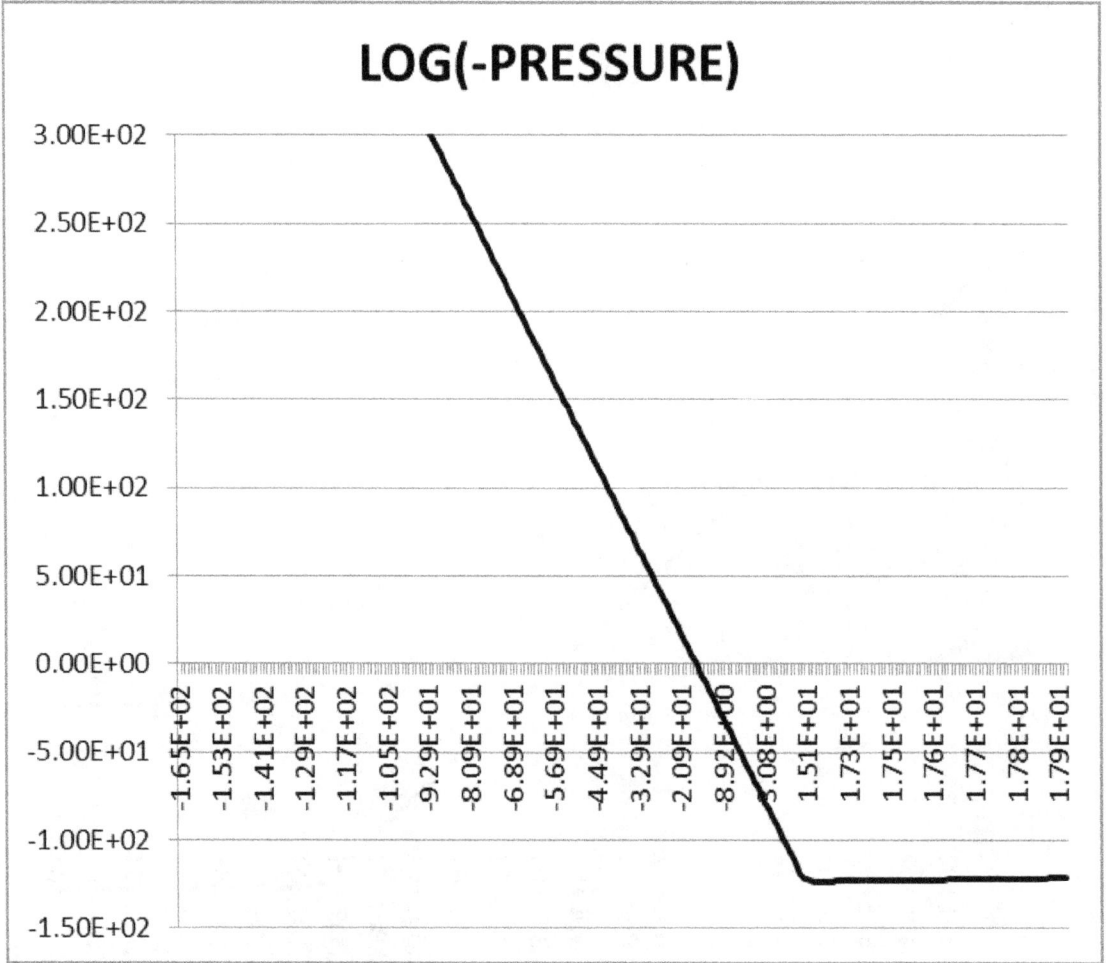

Figure 23.9. Pressure: $\text{Log}_{10}(-p(t))$ plotted vs. $\log_{10}(t)$ from the Big Bang metastate at $t = 1.19 \times 10^{-165}$ sec. to the future time $t = 8.2 \times 10^{17}$ sec. Note the low values of $\log_{10}(-p(t)) \approx -115$ beginning at $\log_{10}(t \text{ sec.}) = 12.1$ ($t = 1.2 \times 10^{12}$ sec.). The pressure is always negative indicating a pressure for expansion. At $t = 1.19 \times 10^{-165}$ sec. we find $\log_{10}(-p(t)) = 593$ or $p(t) = -10^{593}$.

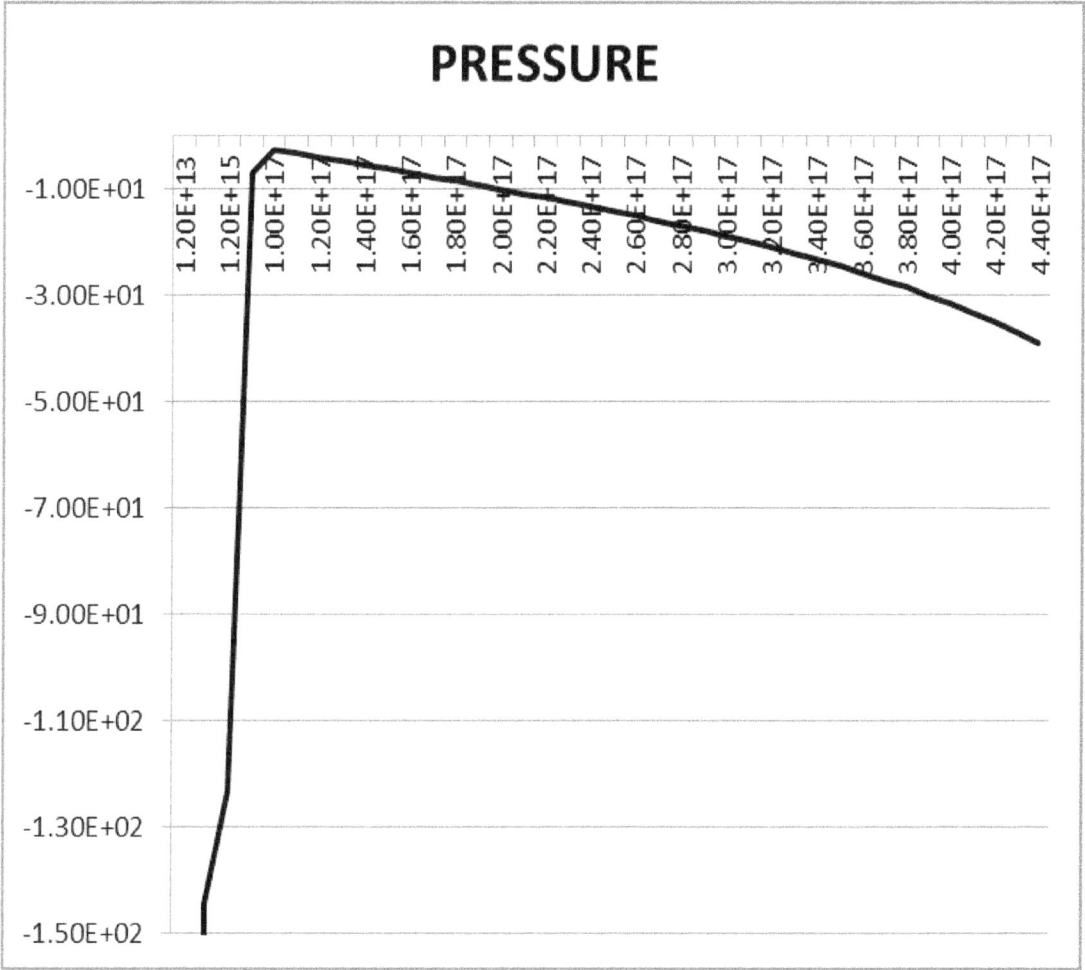

Figure 23.10. Negative Pressure: p(t) $\times$ 10^{124} plotted vs. t sec. from t = 1.19 $\times$ 10^{13} sec. (380,000 years) to the present 4.35 $\times$ 10^{17} sec.

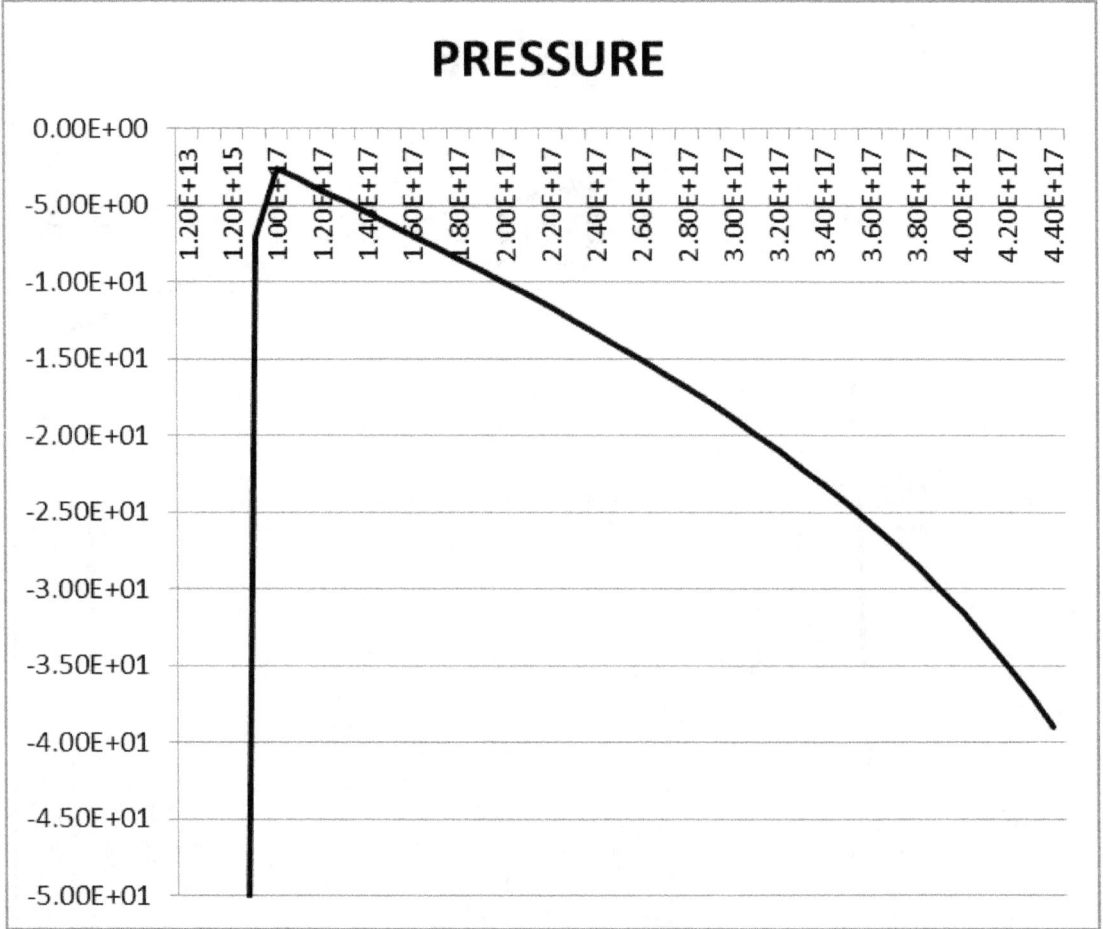

Figure 23.11. Negative Pressure: p(t) $\times$ 10^{124} plotted vs. t sec. from t = 1.19 $\times$ 10^{13} sec. (380,000 years) to the present 4.35 $\times$ 10^{17} sec. The maximum is -2.62 $\times$ 10^{-124} at t = 1.0 $\times$ 10^{17} sec. A closer view than Fig. 23.10.

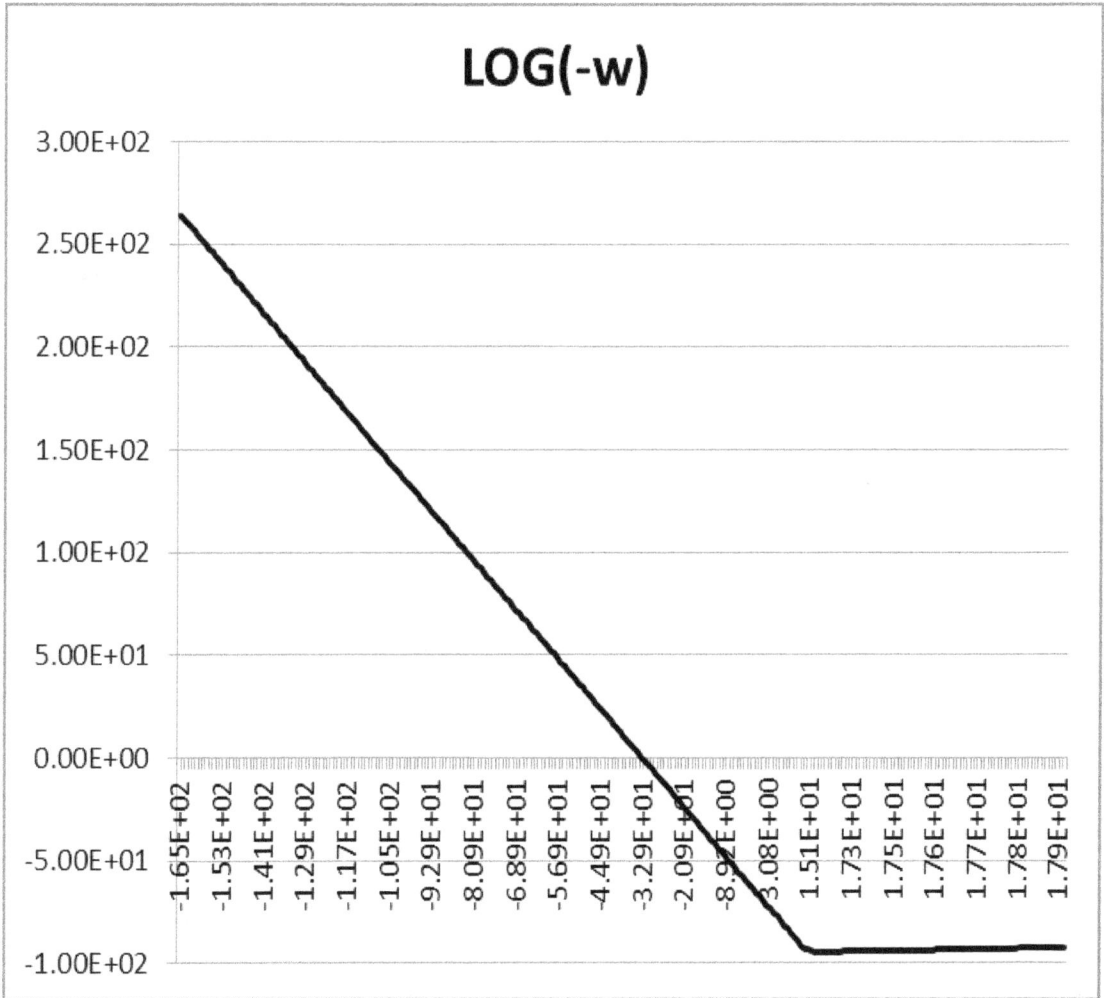

Figure 23.12. The equation of state as a function of time $\log_{10}(-w)$ plotted vs. $\log_{10}(t)$ from the Big Bang metastate at t = 1.19×10^{-165} sec. to the future time t = 8.2×10^{17} sec. Note: w ≠ -1. Quintessence!

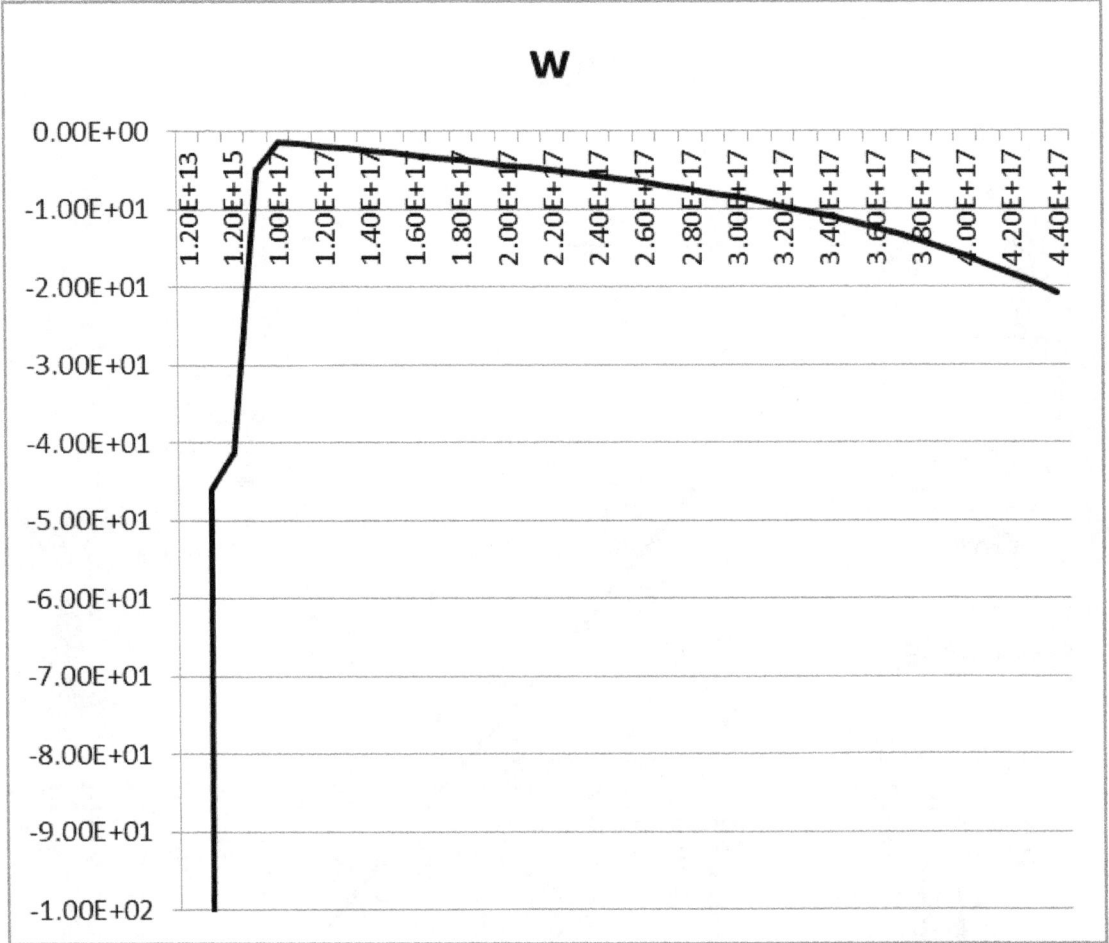

Figure 23.13. The equation of state $w \times 10^{95}$ as a function of time plotted vs. t sec. from $t = 1.19 \times 10^{13}$ sec. (380,000 years) to the present time. Note the peak of $w = -1.36 \times 10^{-95}$ at $t = 1. \times 10^{17}$ sec. Note: $w \neq -1$. Quintessence!

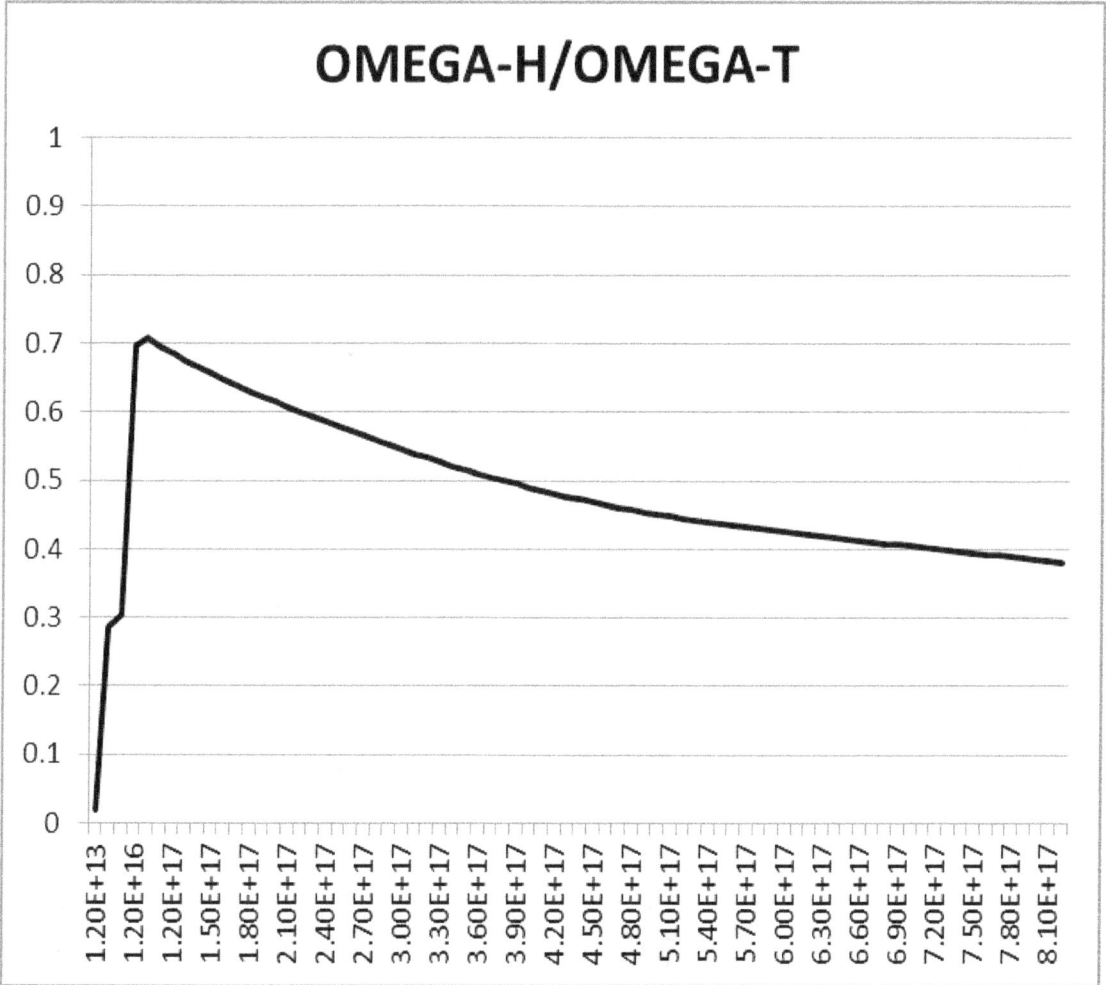

Figure 23.14 The ratio Ω_H/Ω_T showing the dominance of Dark Energy Ω_T from t = 1.19 × 10^{13} sec. (380,000 years) to the future time t = 8.2 × 10^{17} sec. Prior to 380,000 the Dark Energy was greater by many orders of magnitude. The peak value of Ω_H/Ω_T is at t = 1.2 × 10^{16} sec.

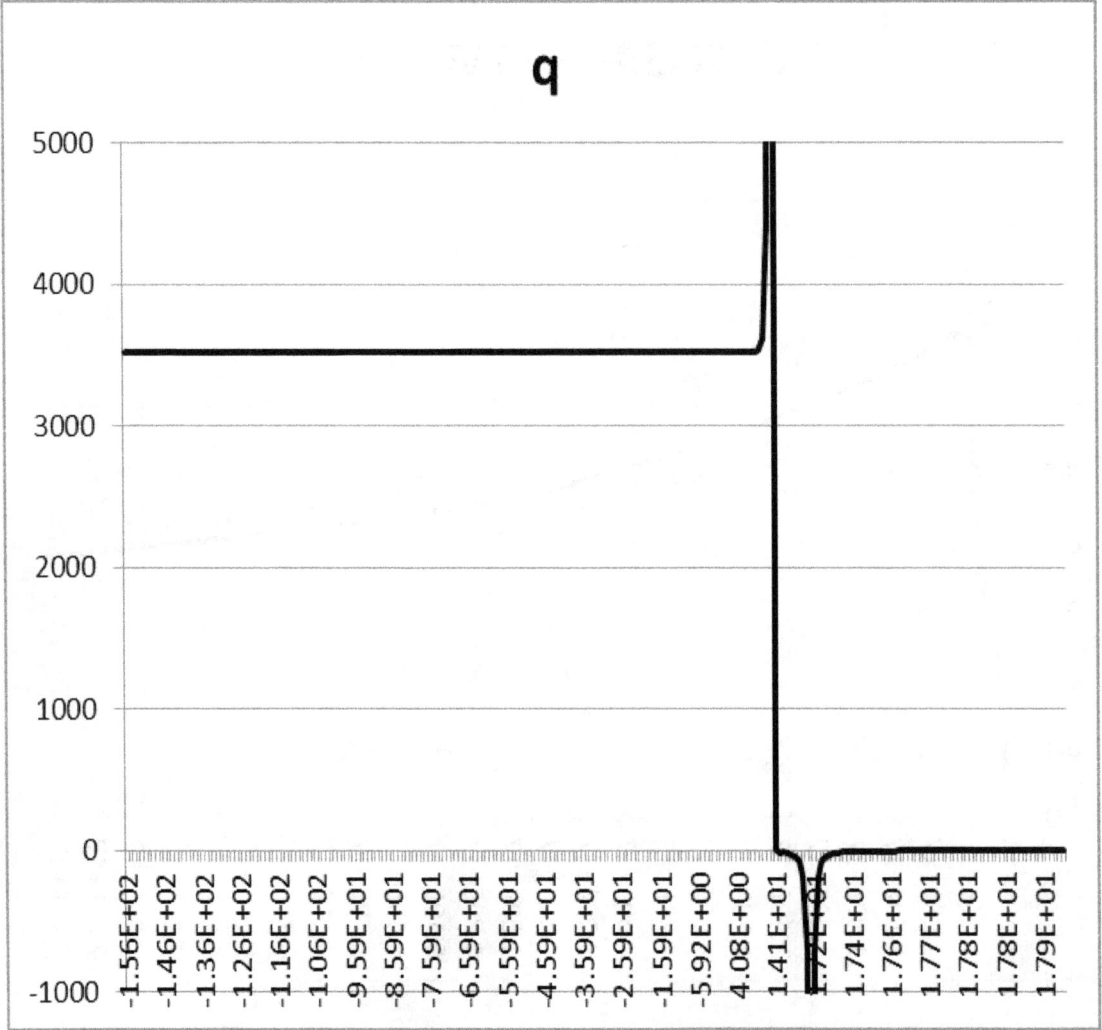

Figure 23.15. The deceleration parameter q vs. log(t sec) from t = 1.2×10^{-156} sec. to the future t = 8.2×10^{17} sec. The peak occurs at t = 1.2×10^{13} sec. The minimum occurs at 1.6×10^{17} sec. Note: q > 0 for t < 1.7×10^{14} sec.

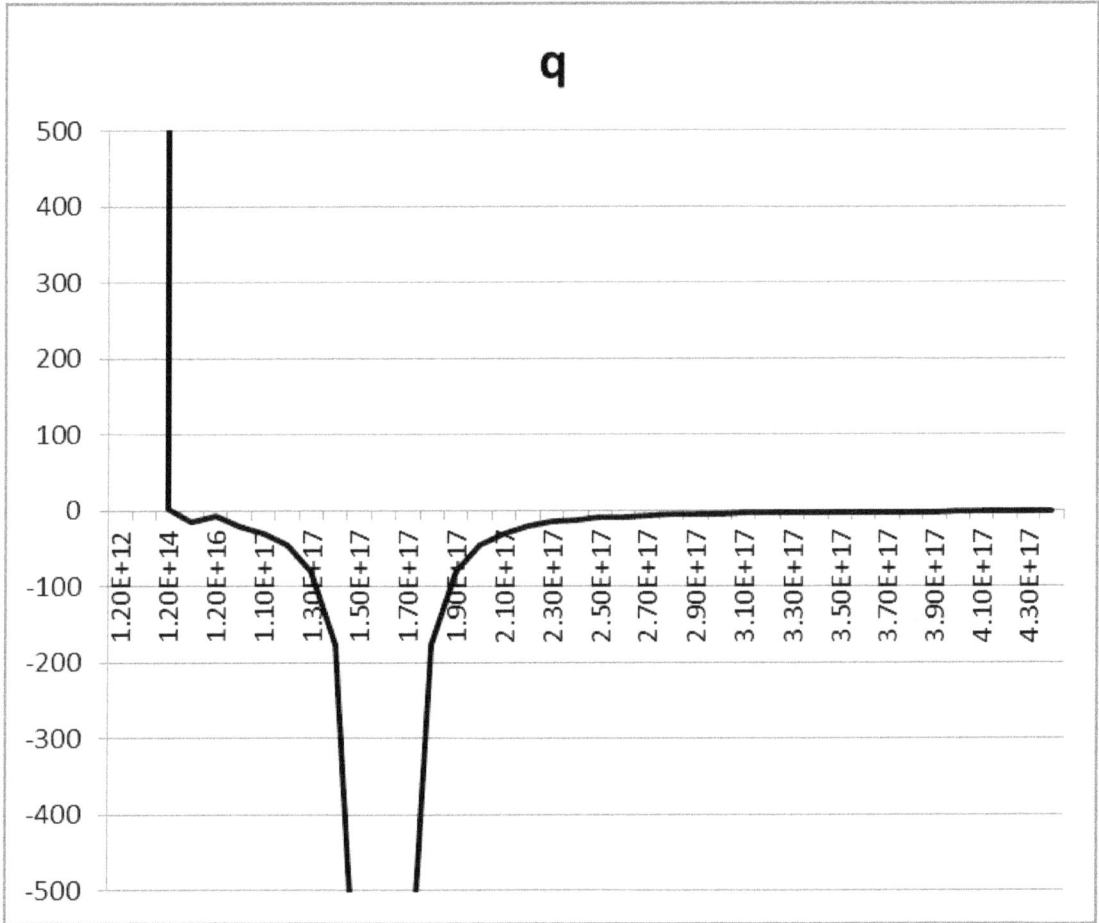

Figure 23.16. The deceleration parameter q vs. log(t sec) from t = 1.2×10^{12} sec. to t = 4.4×10^{17} sec. The peak occurs at t = 1.2×10^{13} sec. The dip occurs at 1.6×10^{17} sec. To the left of the peak q > 0 indicating decelerating expansion. To the right of the peak for t > 1.7×10^{14} sec. we find q < 0 indicating accelerating expansion starting 13.78 billion years ago – "just" after the Big bang. q = -2.0 presently. In the future, at t = 8.2×10^{17} sec. we found q = -1.2.

24. Implications of the Universal Scale Factor

In this chapter we consider some implications of our model Quantum Big Bang and Quantum Vacuum Universe based on the time development of relevant parameters presented in chapters 22 and 23.

24.1 Superclusters and Voids

In our study of the Quantum Big Bang we found that the early Big Bang metastate universe had a wide variation in the Hubble Constant. At its center with radial distance r = 0, we found H = 1.79×10^{218} km s^{-1} Mpc^{-1} while at its periphery H = 1.14×10^{126} km s^{-1} Mpc^{-1} (Fig. 21.1). Thus we see the "explosion" of the energy-mass density of the central region into the outer regions. The result would appear to be a "wave" of mass-energy "emptying" the center and creating a bulge towards the outer region. Thus we anticipate that the primordial universe had variations in mass-energy density favoring the creation of a wave of high mass-energy density and an inner region of low mass-energy. The result is precursors of voids and supercluster bubbles.

In the later Quantum Vacuum Universe period we found a rapid universe contraction to 70% of its previous size (sections 22.6 and 22.7). Subsequently the universe expanded – also fairly rapidly. The contraction (See Fig. 22.6 and the cover of this book) effectively squeezed the outer portions of the universe again creating a wave of mass-energy. The expanding wave then (as in the case of water waves) developed foam (bubbles) of mass-energy of high density that evolved into the superclisters and voids that we find today.

The number of superclusters is estimated to be ten million. Superclusters seem to contain of the order of 10^5 galaxies or more. Thus the combination of the Quantum Big Bang period and the Quantum Vacuum period explain the presence of superclusters and voids found in today's universe.

It appears we have an understanding of the large scale evolution of the universe. Fig. 24.1 presents the large scale pattern of universe evolution.

q rises (Fig. 23.15) to a peak value of 408,273 at t = 1.2×10^{13} sec. indicating decelerating expansion.

Dark Energy Ω_T (Fig. 23.2) rises to a peak of 39.2 at t = 8.71×10^{13} sec. causing H^2 to simultaneously become very large (Einstein equation), although H (Fig. 22.6) is negative and large with a minimum of -445 km s^{-1} Mpc^{-1}.

For t > 1.7×10^{14} sec. we find q < 0 indicating accelerating expansion as shown in H (Fig. 22.6.)

ρ_{tot} declines to a minimum of 1.42×10^{-29} g/cm^{-3} at t = 1.2×10^{16} sec. (Fig. 23.7) after which it rises.

At t = 1.0×10^{17} sec. w and –pressure both peak (Figs. 23.13 and 23.11) Note w $\approx$ 0 suggesting a cold dust or gas. The pressure is also quite small. Both peaks are due to the increase in ρ_{tot} after reaching its minimum.

q > 0 for t < 1.7×10^{14} sec. (decelerating). q reaches a minimum at t = 1.6×10^{17} sec. (Fig. 23.15).

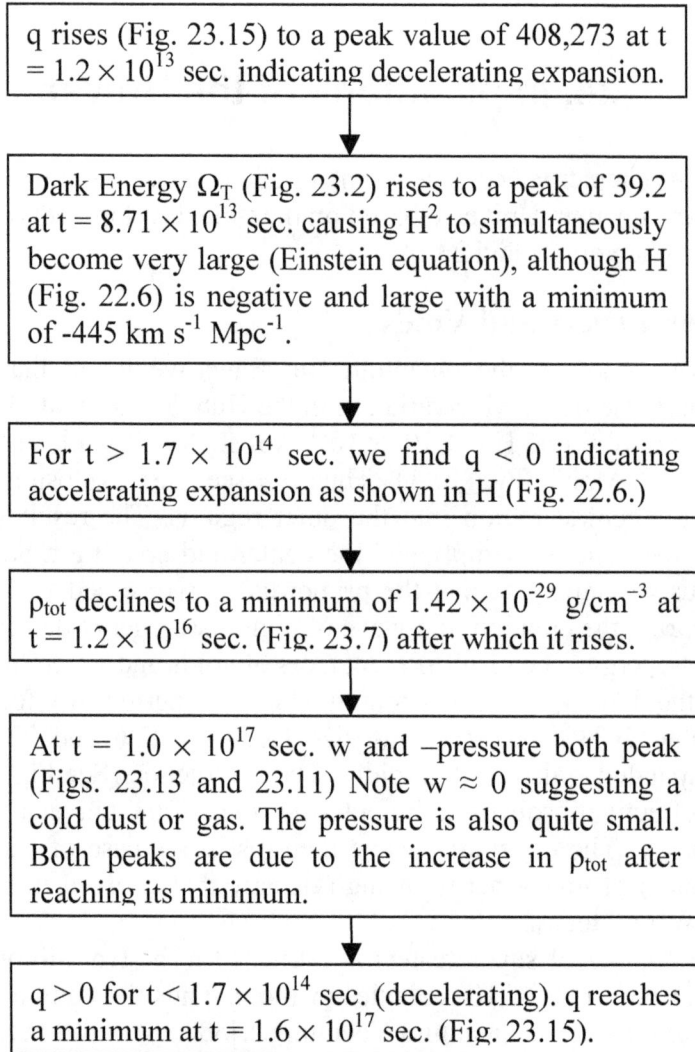

Figure 24.1 Scenario Based on chapters 22-23 figures. The figures in chapter 22 and 23 suggest the evolution in time of the universe.

SECTION 4: PROOF OF UNIVERSAL SCALE FACTOR MODEL

25. Growth of Universes due to Vacuum Polarization

In this chapter[115] we show that the universal scale factor a(t) of eq. 22.1

$$a(t) = (t/t_{now})^{g + ht} \tag{22.1}$$

originates in the vacuum polarization of a spin 0 (boson) universe particle interacting with a new vector QED-like quantum field. Since the vector interaction would presumably affect all universe particles in the Megaverse, it appears that the universal scale factor would apply to all universes. The result would be a common universal pattern of evolution for all universes in the Megaverse.

25.1 Original Motivation for Relating the Scale Factor to Vacuum Polarization

In the preceding chapters it has become evident that the universal scale factor specified by eq. 22.1 leads to a physically reasonable scenario for the life history of a universe. The question that immediately arises is the cause of this form for a(t). How does it work so well for both extremely early times near the Big Bang and for recent times as well?

We begin by expressing the scale factor as

$$a(t, T) = (t/T')^{g + ht} \tag{22.10}$$

where T is a time scale. The value of the exponents g and h are

---

[115] Some of the material in this chapter appears in Blaha (2019c) and (2019d).

$$h = 2.25983 \times 10^{-18} \tag{22.6}$$
$$g = 0.000282377 = 2.82377 \times 10^{-4}$$

For very early times "near" the Big Bang metastate time of $<10^{-165}$ sec we can approximate the universal scale factor with

$$a(t) \cong (t/T)^g \tag{25.1}$$

At first glance the value of g is an arbitrary constant set through the values of the Hubble Constant at t = 380,000 years and T = t_{now}. However the value of g is remarkably similar to the value of a renormalization exponent in the author's calculation of the Fine Structure Constant α in 1973 and in section 5.5.3 of Blaha (2019b).

The value of the Fine Structure Constant was based entirely on vacuum polarization in massless Quantum Electrodynamics (QED) – a vacuum effect of electromagnetism. Although astrophysicists do not think of the Big Bang and the expansion of the universe as a vacuum effect, it is clear from the plots shown earlier that universe growth is dependent on its energy density and an influx (appearance) of energy from somewhere (the Megaverse in our view). Thus the growth of the universe is directly dependent on the vacuum (of the Megaverse).

In massless QED we found that the vacuum polarization had the form:[116]

$$F_1(\alpha)(p/\Lambda)^{2g_{QED}} \tag{25.2}$$

where $F_1(\alpha)$ is the "eigenvalue function" for the Fine Structure Constant[117] of the Johnson-Baker-Willey model of massless QED, p is the momentum, and Λ is the ultraviolet cutoff. The value of g_{QED} that corresponded to the Fine Structure Constant is[118]

---

[116] Eq. 12 in S. Blaha, Phys Rev **D9**, 2246 (1973).
[117] The author calculated α = 1/137… exactly (within experimental limits) in Blaha (2019a) and (2019b).
[118] Section 5.5.3 of Blaha (2019b) had an infinitesimally different value which is herein corrected.

$$g_{QED} = -0.0005805375 \qquad (25.3)$$

and the Fine Structure Constant was correctly found (well within experimental limits) to be

$$\alpha_{calculated}(g_{QED}) = 0.007297353 \qquad (25.4)$$

to 9 digit accuracy according to the Particle Data Table of 2018.
Comparing our g value (eq. 22.6 above) with g_{QED} we

$$-g \cong -\tfrac{1}{2} \, g_{QED} \qquad (25.5)$$

Excepting the factor of two, we find a remarkable numerical coincidence comparing electron vacuum polarization with the universe scale factor where high momentum electron polarization corresponds to very early universe time. This coincidence is not accidental.

25.2 A New Vector Interaction for Universe Particles

We assume universes can be treated as particles in 4-dimensional space-time.[119] Since experiments appear to have shown that the universe does not rotate (does not have spin)[120] we will assume the universe is a spin 0 boson. We assume that universes have a vector field interaction similar to QED. It is possible that the quantum vector $Y^{\mu}(x)$ field of the Big Bang quantum coordinates, treated earlier, may be the vector field universe interaction field as well.

Given this QED-like framework, universe-antiuniverse pair production and vacuum polarization becomes possible. We assume the QRD-like lagrangian

$$\mathcal{L} = \tfrac{1}{2}\,(\partial_{\mu}\varphi^{\dagger}\partial^{\mu}\varphi - m^2\varphi^{\dagger}\varphi) - ie_0 \!: \varphi^{\dagger}(\overrightarrow{\partial_{\mu}} - \overleftarrow{\partial_{\mu}})\,\varphi : A^{\mu} + e_0^{\,2}\!:\!A^2\!: \!:\!\varphi^{\dagger}\varphi\!: + \delta m^2\!:\!\varphi^{\dagger}\varphi\!:$$

$$(25.6)$$

---

[119] Universes are composite entities but we can treat them as quantum particles in the same manner as physicists treated protons and neutrons etc. as quantum particles before quark theory was accepted.

[120] The lack of universe rotation (spin) is indicated by a study of Cosmic Microwave Background (CMB) by D. Saadeh *et al*, Phys. Rev. Lett. **117**, 313302 (2016).

where $\varphi(x)$ is a "charged" quantum universe particle field.[121]

We now proceed to calculate the second order vacuum polarization of a universe particle.

25.3 Second Order Vacuum Polarization of a Universe Particle

The one loop vacuum polarization Feynman diagram is

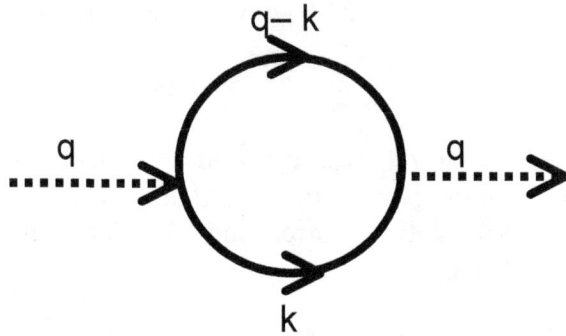

Figure 25.1 One loop vacuum polarization boson Feynman diagram.

We evaluate

$$I_{\mu\nu} = (-ie_0)^2 \int \frac{d^4k}{(2\pi)^4} \; \frac{i}{(k^2 - m^2 + i\varepsilon)} \; \frac{i}{(k^2 - m^2 + i\varepsilon)} (q - 2k)_\mu (q - 2k)_\nu \qquad (25.7)$$

$$= \frac{\alpha}{2\pi} \int_0^\infty dz_1 \int_0^\infty dz_2 \frac{g_{\mu\nu}}{(z_1 + z_2)^3} \exp[i(q^2 z_1 z_2/(z_1 + z_2) - (m^2 + i\varepsilon)(z_1 + z_2))] + \text{gauge terms}$$

upon introducing parameters z_1 and z_2 to enable exponentiation and integration over k, where

---

[121] The charge is not electromagnetic charge.

$$\alpha = e_0^2/4\pi \tag{25.8}$$

Applying $q^2 \partial/\partial q^2$ to $I_{\mu\nu}$ to eliminate the quadratic divergent part, and then using the identity

$$1 = \int_0^\infty d\lambda/\lambda \; \delta(1 - (z_1 + z_2)/\lambda)$$

and letting $z_i = \lambda x_i$ we obtain

$$I_{\mu\nu} = \frac{i\,\alpha}{2\pi} \; q^2 g_{\mu\nu} \int dx_1 \int dx_2 \int d\lambda/\lambda \; x_1 x_2 \exp[i\lambda(q^2 x_1 x_2 - (m^2 + i\varepsilon))] \; \delta(1 - x_1 - x_2)$$

$$\tag{25.9}$$

up to gauge terms. The λ integration yields a logarithmic divergence which we cut off. Then

$$I_{\mu\nu} = \frac{i\,\alpha}{2\pi} \; q^2 g_{\mu\nu} \int_0^1 dx \; x(1\text{-}x) \ln(q^2 x(1 - x) - m^2) \tag{25.10}$$

which becomes

$$I_{\mu\nu} = \frac{i\,\alpha}{12\pi} \; q^2 g_{\mu\nu} \ln(\Lambda^2/m^2) + \ldots \tag{25.11}$$

with finite terms not shown.

Thus we find the renormalization constant Z_3 for the scalar universe particle case to be

$$Z_3 = 1 - \alpha/12\pi \; \ln(\Lambda^2/m^2) \tag{25.12}$$

If we let

$$\alpha_U = \alpha/4 \tag{25.13}$$

then we obtain the form similar to the one loop value of Z_3 for spin ½ electron QED:

$$Z_3 = 1 - \alpha_U/3\pi \; \ln(\Lambda^2/m^2) \tag{25.14}$$

We now provisionally assume that α is the QED fine structure constant. We denote it as α_{QED}. Note $\alpha_{QED} = 0.007297353$. (Later we will show that the value of $g = 0.000282377$ that we found earlier follows from this choice.)

Thus the "fine structure constant" α_U for our vector interaction is

$$\alpha_U \equiv \alpha_{QED}/4 = 0.001824338 \tag{25.15}$$

We now turn to the Johnson-Baker-Willey (JBW) model of massless QED since at ultra high energy our vector interaction theory with lagrangian eq. 25.6 becomes the JBW model. In the JBW model we calculated α_{QED} and found the corresponding power which we denote g_{QED}. Now we perform the same calculation and find the g value denoted g_U corresponding to α_U. The value of g_U will be seen to lead to the power g in the universal scale factor almost exactly.

25.4 Massless QED-like Calculation of Vacuum Polarization to All Orders

We calculated[122] an approximate solution to all orders of the four divergences in the JBW model:

$$
\begin{array}{ll}
Z_1 & \text{- vertex renormalization factor} \\
Z_2 = Z_1 & \text{- self-energy renormalization factor} \\
Z_3 & \text{- vacuum polarization renormalization factor} \\
\delta m & \text{- self-mass renormalization}
\end{array}
$$

The renormalization constants appear in the expressions:

$$
\begin{array}{l}
S_F(p) = Z_2 S'_F(p) \\
D_F(q)_{\mu\nu} = Z_3 D'_F(q)_{\mu\nu} \\
V_\mu(q', q) = Z_1^{-1} V'_\mu(q', q)
\end{array}
$$

---

[122] See S. Blaha, Phys. Rev. **D9**, 2246 (1974). Reprinted in Appendix F. Blaha (2019b) presents the JBW model and the calculation of α_{QED}.

where $S'_F(p)$, $D'_F(q)_{\mu\nu}$, and $V'_\mu(q', q)$ are the divergence-free QED physical propagators and vertex.

The JBW model was an attempt to eliminate the divergences of QED at very high energies where the electron mass may be neglected. It was extended to allow for non-zero electron mass.

The massless QED eigenvalue function of the JBW model was found in a series of papers and summarized in some detail by the paper of K. Johnson and M. Baker, Phys. Rev. **D8**, 1110 (1973) In this section we briefly outline the steps leading to the JBW eigenvalue function:

1. The electron self-energy and the photon vacuum polarization are calculated using the free electron and photon propagators.

2. The apparent quadratic divergence is reduced to a logarithmic divergence in the vacuum polarization.

3. The logarithmic divergence in the vacuum polarization is further reduced to a single power of the logarithm of the ultraviolet cutoff denoted Λ.

4. The coefficient of the divergent logarithmic term is denoted

$$(x/2\pi)F(x)$$

where x is the bare fine structure constant α_0.

5. If

$$F(x_0) = 0$$

for some value x_0 then a consistent divergence-free (finite) solution of massless QED is found.

6. The F(x) function is reduced to the eigenvalue function $F_1(x)$ which is the sum of logarithmic divergences of all single closed electron loop diagrams. If the eigenvalue function $F_1(x)$ has a zero at x_0 then $F(x_0) = 0$. Consequently the eigenvalue condition becomes

$$F_1(x_0) = 0$$

7. Adler[123] made the important observation that a zero F_1 would necessarily be an essential singularity:

$$d^n F_1(x)/dx^n|_{x=x_0} = 0 \qquad \text{for all } n > 0$$

Depending on the summation in perturbation theory of the relevant vacuum polarization diagrams might may occur at the bare coupling constant α_0 or the physical coupling constant $\alpha = 1/137 \dots$

8. This author (See Appendix F.) calculated an approximation to F_1 which did not explicitly display an essential singularity and did not have a zero at the physical fine structure constant α.

25.5 Blaha's Approximate Calculation of the Eigenvalue Function

In 1974 this author[124] formulated an approximation to the equations of massless QED and solved them for the vacuum polarization, electron self-energy and the vertex renormalization. The approximation is described in detail in the author's Phys. Rev. D paper in Appendix F.

The approximate solution for $F_1(x)$ had the encouraging feature that it reproduced the known[125] low order exact calculations of $F_1(x)$:

$$F_{1 \text{ low order}}(x) = 2/3 + x/(2\pi) - (1/4)[x/(2\pi)]^2$$

[123] Adler *op. cit.*
[124] Blaha *op. cit.*
[125] J. Rosner, Phys. Rev. Lett. **17**, 1190 (1966).

The *approximate* solution, which summed pieces of the vacuum polarization given by the diagrams of Figs. 2 and 3 in Appendix F, yields the algebraic equations:[126]

$$A_1 = (g + 1)(1 - 2g^2)/[(g + 2)(g - 1)]$$

$$A_2 = [8g^2(2g + 1) - (2g^3 + 2g^2 + g - 2)(g^2 + 2g + 2)]/[2(g^2 - 1)(g^2 - 4)]$$

$$A_3 = -2(1 + 3g + 6g^2 + 2g^3)/[g(g + 1)]$$

$$A_4 = -(g + 2)(1 + 5g + 6g^2 + 2g^3)/[g(g^2 - 1)] - 1/(g + 1)$$

$$\psi = [gA_3 - (4 + 2g)A_1]/[(4 + 2g)A_2 - g A_4]$$

$$(\alpha/2\pi) = [gA_4 - (4 + 2g)A_2]/(A_4A_1 - A_2A_3) \tag{25.16}$$

$$F_1(g) = (2/3)(1 - 3g^2/2 - g^3) - (\alpha/4\pi)[(2 + 4g + 4g^2)(g - 2) + \alpha\psi g^3]/[(g^2 - 1)(g - 2) + \\ + \alpha(2 + 4g + 4g^2)(g - 2) + \alpha\psi g^3]$$

as a function[127] of $g = g_{QED}$ with ψ specifying the gauge, and with the definitions

$$\Gamma_\mu(p) = f(\gamma_\mu + 2g\gamma \cdot pp_\mu/p^2)(p/\Lambda)^{2g} \tag{25.16'}$$
$$S_F = [f\gamma \cdot p(p/\Lambda)^{2g}]^{-1}$$
$$\Gamma_{\mu\alpha}(p) = (f_3/p^2)(\gamma \cdot p\gamma_\mu\gamma_\alpha - \gamma_\alpha\gamma_\mu\gamma \cdot p)(p/\Lambda)^{2g}$$

and

$$F_1 = (2/3)(1 - 3g^2/2 - g^3) - f_3/f$$

in the notation of Appendix F. Eqs. 5.4 and 5.5 manifestly cannot lead to a form of F_1 with an essential singularity due to their algebraic form.

---

[126] Blaha *op. cit.* The solution for the eigenvalue function is clearly best expressed in terms of the g factor in the exponents of the divergent renormalization factors.
[127] We use $F_1(g)$ and $F_1(\alpha(g))$ interchangeably.

The plot of F_1 did not show a zero of F_1 at the physical fine structure constant. Thus the hopes raised by the JBW model seemed dashed—at least in our approximate solution *then*. Later in Blaha (2019a) and (2019b) we revived the hope of a satisfactory eigenvalue function with an eigenvalue at the physical value of the fine structure constant α.

As pointed out in our 1974 Phys. Rev. D paper the eigenvalue function F_1 does not have a zero at the known value of the Fine Structure Constant.

In 2019 we reconsidered our approximation and found a value of α very near the measured value. We defined

$$F_2(\alpha) = F_1(\alpha) - [2/3 + \alpha/(2\pi) - (1/4)[\alpha/(2\pi)]^2] \qquad (25.16'')$$

25.6 The *Correct* Value of the QED Fine Structure Constant

We found a neighborhood in the range of g where $F_2(\alpha) \approx 0$ with an approximate value for the known fine structure constant. We reconsidered our calculation recently and found

$$g_{QED} = -0.000580537 \qquad (25.17)$$

and

$$\alpha_{QED} = 0.007297353 \qquad (25.18)$$

which gives the QED fine structure constant (an irrational number) accurately to 9 places.

25.7 Extension of Calculation to Non-Abelian Interactions

In Blaha (2019b) we generalized eq. 25.16 to the cases of the Weak interaction and the Strong interaction coupling constants by inserting a group theoretic factor in eq. 25.16:

$$c_G^{-1} = [(11/3)C_{ad} - 2C_f/3]/(16\pi)^3 \qquad (25.19)$$

$$(\alpha_G/2\pi) = c_G^{-1}[gA_4 - (4 + 2g)A_2]/(A_4A_1 - A_2A_3) \qquad (25.20)$$

where C_{ad} is the dimension of the fundamental representation of the non-abelian group and C_f is the number of fermions (fermion flavor) of the interaction.

We found good approximations to the SU(2) and SU(3) coupling constants.

25.8 Application of the Approximate Coupling Constant Calculation to Universe Vacuum Polarization

We now extend eq. 25.16 to the case of the vacuum polarization of a universe particle described by

$$(\alpha_U/2\pi) = [g_U A_4(g_U) - (4 + 2g_U)A_2(g_U)]/(A_4(g_U)A_1(g_U) - A_2(g_U)A_3(g_U)) \quad (25.21)$$

since the second order form of Z_3 (eq. 25.14), which is the same as the QED second order form of Z_3, generalizes to all orders as a function of α_U.

Given the value of

$$\alpha_U = 0.001824338 \quad (25.22a)$$

in eq. 25.15 we can extract the value of g_U by inverting eq. 25.21 to obtain:

$$g_U = -0.00014525 \quad (25.22b)$$

25.9 Relating Vacuum Polarization to the Universal Scale Factor

We now relate the vacuum polarization found above to the growth of the universe as given by the universal scale factor. Eq. 25.16'

$$\Gamma_{\mu\alpha}(p) = (f_3/p^2)(\gamma \cdot p\gamma_\mu\gamma_\alpha - \gamma_\alpha\gamma_\mu\gamma \cdot p)(p/\Lambda)^{2g_U}$$

gives the vacuum polarization factor

$$\Gamma(p) = (p/\Lambda)^{2g_U} \quad (25.23)$$

where

$$g_U = g/4 = -0.00014543 \quad (25.24)$$

We now fourier transform $\Gamma(p)$ to coordinate space – in particular to time t[128]

$$a(t) = (1/2\pi) \int_0^\infty dp \, \exp(-ipt) \, \Gamma(p) \qquad (25.25)$$

$$= k \, (t/T)^{-2g_U} \qquad (25.26)$$

where k is a constant and where

$$1/T = \Lambda \qquad (25.27)$$

with Λ being the "momentum space" cutoff mass. From eq. 25.22b and 25.26 we find

$$g = -2g_U$$
$$= 0.0002905 \qquad (25.28)$$

From eq. 22.6 for the power g of a(t) we see

$$g = 0.000282377$$

Thus the value of g calculated from the vacuum polarization differs from the actual value of g by less than 3%. Given the approximate nature of our JBW calculation of vacuum polarization the agreement is remarkable.[129]

The dependence of the universal scale factor on g governs the small time behavior of the universe. Correspondingly, the dependence of the vacuum polarization on g_U is a large momentum phenomena. The parameter h in a(t) is set primarily by the large t (recent times) behavior of a(t). It corresponds to the infrared behavior of its

---

[128] Those who might object to forier transforming to time t should remember that inside a Black Hole the "time-like" coordinate is the radius and the time variable t is comparable to a spatial coordinate. The possibility that the universe is a Black Hole is not excluded.

[129] And may be exact! The value of the Hubble Constant H in recent times varies from about 70 – 75 making the calculation of g also approximate. We chose an average value of 73.24 to obtain the value of g above. If we chose the current value for H to be 75.58 we would have $g = -2g_U$ exactly (eq. 25.28). Note: studies of binary black hole merger gravity waves have given a Hubble Constant of 75.2 km s^{-1} Mpc^{-1} (and earlier of 78 km s^{-1} Mpc^{-1}), and studies of light bent by distant galaxies give H = 72.5 km s^{-1} Mpc^{-1}. Thus the value H = 75.58 is not unreasonable. See section 22.1 for a summary of studies of H.

fourier transform $\Gamma(p)$ when the infrared (possibly mass dependent) behavior of the vacuum polarization is calculated.

The preceding discussion demonstrates that our earlier assumption

$$\alpha_U \equiv \alpha_{QED}/4 = 0.001824338 \qquad (25.15)$$

is correct since it leads directly to the power g in the universal scale factor. Thus the evolution of our universe is set by the vacuum polarization originating in the new vector interaction that we have introduced.

Using our notation we can rewrite eq. 25.6 as

$$\mathcal{L} = \tfrac{1}{2}\,(\partial_\mu\varphi^\dagger\partial^\mu\varphi - m^2\varphi^\dagger\varphi) - ie_{0U}: \varphi^\dagger(\overrightarrow{\partial_\mu} - \overleftarrow{\partial_\mu})\,\varphi: Y^\mu + e_{0U}{}^2:Y^2: :\varphi^\dagger\varphi: + \delta m^2:\varphi^\dagger\varphi:$$

$$(25.29)$$

The Y^μ vector field that we used to stabilize the Big Bang with quantum coordinates may now have a role as the interaction between universe particles. Incidentally we have also shown the viability of viewing universes as quantum particles just as nucleons were viewed as quantum particles before the acceptance of a quark substructure.

25.10 Proof that Our Universe's Growth Pattern is due to a Form of Vacuum Polarization

The calculation of g from the vacuum polarization of a vector field proves that the growth pattern of our universe is governed by vacuum polarization although it takes the form of depending on Dark Energy. It appears that Dark Energy *per se* does not exist except as a representation of the vacuum polarization generated by our vector gauge field theory. Note no direct physical evidence of Dark Energy interacting with "normal" matter exists. Dark Energy is inferred and may be vacuum polarization energy due to a new universe interaction.

Since the universal scale factor implies enormous, varying Dark Energy Ω_T as shown in chapter 23, we view the growth of our universe as a vacuum polarization phenomenon.

25.11 Vacuum Polarization Interpretation of the Universal Scale Factor

The vacuum polarization view of the time evolution of the universe requires that we view the entire time evolution of the universe as a whole. Normally one views time as increasing. Feynman suggested we could also view time as flowing backward.

We now have a new view where we freeze the life history of the universe as a static event rather like a time lapse picture of a flower's growth. The thought process is similar to that of Feynman path integral formulations, which consider the complete path in time of a process.

25.12 Path Integral Formulation of a *Quantum* Scale Factor

It is possible define a path integral formalism for a *quantum* a(t) starting from the Friedmann equation and its lagrangian formulation. The Friedmann equation

$$d^2a/dt^2 + 4\pi G/3 \ (\rho + 3p/c^2) \ a - \Lambda c^2 a/3 = 0$$

clearly resembles the Schrödinger equation. The sum over paths of a(t) would then have a path corresponding to the "classical" solution eq. 22.1 that we considered. Thus the complete "history" in a(t) (and its vacuum polarization equivalence) would be understandable. A potential benefit of the quantum a(t) is the elimination of divergences at t = 0 by quantum smearing.

25.13 Scale Factors of Other Universes

The considerations of a vector interaction generating the universal scale factor of our universe also applies to other universes in the Megaverse due to the assumption of a common Y interaction for all universes. Thus the growth pattern of our universe applies to other universes. It is a general feature of universes.

Other universes have the same evolutionary development as our universe.

25.14 Universe Eigenvalue Function

The eigenvalue function for universes is based on the general eigenvalue function

$$F_2(\alpha) = F_1(\alpha) - [2/3 + \alpha/(2\pi) - (1/4)[\alpha/(2\pi)]^2] \qquad (25.16'')$$

The universe eigenvalue function is

$$F_2(\alpha_U) = F_1(\alpha_U) - [2/3 + \alpha_U/(2\pi) - (1/4)[\alpha_U/(2\pi)]^2] \qquad (25.30)$$

If

$$F_2(\alpha_U) = 0 \qquad (25.31)$$

as it does for

$$\alpha_U \equiv \alpha_{QED}/4 = 0.001824338 \qquad (25.15)$$

where we find

$$F_2(\alpha_U = 0.001824338) = 5.10824 \times 10^{-12}$$

then, given the approximation used, $F_2(\alpha_U = 0.001824338)$ is essentially zero.

Just as a zero of F_2 implies quasi-free particle behavior at high energy in the JBW model we find universe particles are quasi-free.[130] (Gravitation between universes still exists.) See Appendix D for the case of free spin ½ universes.

25.15 A Hierarchy of Interactions

The vector interaction which we denote with the quantum field label Y with $e_U = (4\pi\alpha_U)^{1/2}$, and the other coupling constants for QED, Weak SU(2) and Strong SU(3) have a remarkable regularity—they double from interaction to interaction as Fig. 25.2 shows.

The deeper significance of this regularity is not known.

---

[130] Universe particles do have a low order divergent piece (exhibited in eq. 25.16″) in their vacuum polarization as shown in Blaha (2019b).

INTERACTION	COUPLING CONSTANT[131]
Y Interaction e_U	0.152
QED $e_{QED} = (4\pi\alpha_{QED})^{1/2}$	0.303
Weak SU(2) g_W	0.619
Strong SU(3) g_S	1. 22

Figure 25.2. The interaction constants show a regular doubling. The cause of the doubling is not apparent.

[131] M. Tanabashi *et al* (Particle Data Group), Phys. Rev. D**98**, 030001 (2018).

26. Universe Particle Dynamics

The origin of our universe and other universes is a weighty question that has been considered in a number of theories. Our view is that the existence of the Megaverse, and the locality of universes (in the one case of which we are certain), suggest the Megaverse is the "platform" of a type of particle physics in which universes play the role of particles. There are clear points of difference: particles have a fixed mass, particles have well-defined quantum numbers, and so on. Nevertheless we can consider universes to be variable mass particles, *universe particles*, with a dynamics that is primarily based on gravitation.

This chapter describes aspects of the interactions of universe particles due to gravitation, baryon number forces, and collisions between universes. It also the describes the genesis of universes due to vacuum fluctuations, the fission of universes, and the internal distortion of universes due to acceleration and the presence of 'nearby' universes..

26.1 The Internal Distortion of Universes

In the absence of external forces universes are considered to be uniform in the large. However the acceleration of a universe in the Megaverse can distort the universe. Also the existence of a nearby universe(s) could cause the uniformity of a universe to be lost due to gravitation and baryonic forces.[132]

26.1.1 Impact of Universe Particle Acceleration – Lopsided Internal Structure of Universe

Universes can accelerate within the Megaverse due to external Megaverse forces. Universe acceleration should be detectable within a universe as "lopsidedness."

---

[132] The baryonic forces, the Baryonic force and the Dark Baryonic force, on a universe are large due to their additivity in our universe and nearby universes.

There would be a shift of parts of the universe opposite to the direction of acceleration resulting in a difference in the features of the universe "in front" compared to those "in back" – an acceleration effect just as one sees when a jet accelerates.

Interestingly new data from the Planck observatory of the European Space Agency confirms and extends earlier data from NASA's WMAP observatory that one side of the universe appears different from the other side. There are temperature differences and mass distribution differences – just as one might expect if the universe were accelerating as a unit.

Thus we see the beginning of data suggesting our universe may be accelerating through the Megaverse. Some Planck observatory scientists have suggested their data is a preliminary indication of the Megaverse.

26.1.2 Impact of External Forces on Universe Structure

The presence of a nearby universe could cause a universe to lose its large scale uniformity and the mass-energy of the universe to drift over time to the 'nearby' side of the universe due to gravitation and baryonic forces.

26.2 Universes in Collision

We can assume that the dynamics of universes in collision will be analogous to that of galaxies in collision since gravity is a dominant force in both cases. Colliding galaxies have often been observed. Their dynamics should provide guidance for the case of universes in collision.[133]

It is clear in the case of colliding galaxies, and of colliding large nuclei (gold and lead typically) that there are several types of collisions with differing results. Similarly, the types of universe collisions can be qualitatively classified as:

5. Clean collisions in which universes nudge each other but retain their identity. These are extreme peripheral collisions. If the universes overlap slightly then the

---

[133] The high energy collision of atomic nuclei at Brookhaven, CERN and other laboratories also is analogous in overall detail with universes in collision.

typically spherical symmetry of the universes may become distorted and they may become lopsided.[134]

6. Peripheral collisions in which the universes retain their identity but are connected by a trailing string of mass-energy. Eventually the string breaks and the universes separate. Subsequently the pieces of trailing string in each universe contract due to their universe's gravitational effects and perhaps form new "bubble" universes.

7. Two universes can collide and produce multiple universes.

8. Two universes can collide in a "central" collision and amalgamate into one universe. They can intermix with both the baryonic, gauge, and gravitational forces causing a redistribution of their masses. They may separate afterwards or may coalesce into a single universe. One result of this may be lopsided universes. Our universe appears to be lopsided. Some cosmologists believe this is due to a near collision of our universe with another shortly after the Big Bang.

26.3 Creation of Universes through Gauge Field Fluctuations

One of the most exciting questions in Cosmology is the origin of our universe. The conventional view is that it originated in a Big Bang from an infinitesimal point in space. The source of the Big Bang and the prior state of the Megaverse, if there was one, is the subject of much speculation. Based on the particle interpretation of the Wheeler-DeWitt equation, the possibility of a baryonic force strongly supported by conservation of baryon number, and the Megaverse concept, it is reasonable to consider the possibility that the universe originated in a vacuum fluctuation.

In this case there would be two Big Bangs one for our universe and one for an anti-universe. One would expect that they would have opposite corresponding features:

---

[134] The Wilkinson Microwave Anisotropy Probe (WMAP) and the Planck European Space Agency satellite have been accumulating data since 2001 that suggests the universe may be lopsided with hot and cold spots on opposite sides of the universe differing from those on the other side being hotter and colder respectively—*perhaps the result of a collision when the universe was young.*

one with baryon dominance – one with anti-baryon dominance, and one left-handed – one right-handed.

Our formulation of universe particle theory allows for the generation of a universe particle and anti-particle as a vacuum fluctuation. We view a universe particle as having a substantial excess of baryons, N, as we see in our universe. Its anti-universe at the time of creation (the Big Bang point) is its "mirror image" having the "same" number of anti-baryons (baryon number –N) so that baryon number is conserved by the fluctuation event. Thus the excesses of one universe are compensated by the excesses of the other.

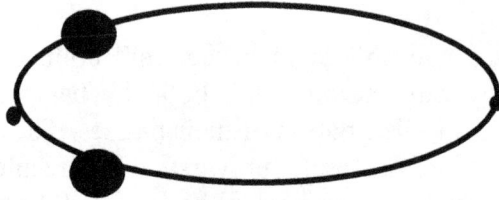

Figure 26.1. Generation of a universe – anti-universe pair as a vacuum fluctuation. The resemblance to the vacuum polarization diagram of chapter 25 is clear.

The small value of the coupling constant should lead to an extremely long lifetime for the universes generated by the fluctuation. Thus the 45 billion year life of our universe is not unreasonable. The probability of the creation of universes by vacuum fluctuations should be correspondingly small.

The sizes of the created universe and antiuniverse should be very small just as Big Bang theories of our universe suppose.

26.4 Fission of Universes

Under certain circumstances the distribution of matter in the universe may lead to the fission of the universe into two separate universes. Our theory supports this possibility for universe particles. The detailed mechanism of the fission process is not specified by the model.

26.4.1 Fission of Normal Universes

The fission of universe particles in our universe particle model is depicted in the Feynman diagram in Fig. 26.2.

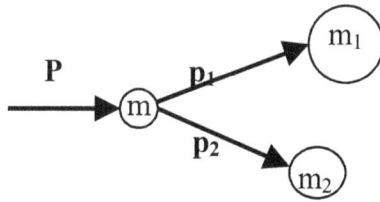

Figure 26.2. Fission of a universe particle into two universe particles.

The sum of the masses of the output universe particles is usually less than the original universe particle mass. However if the fission takes a long time and the masses are time dependent then the produced universe particles combined masses may exceed the original universe's mass.

26.4.2 Tachyon Universe Particle Fission to More Massive Universe Particles

In Blaha (2007a) we showed that a tachyon (faster than light) particle could fission into particles of larger mass through the conversion of momentum into mass. In this section we show that a tachyon universe particle may fission into two more massive universe particles.[135] This phenomenon is of particular interest because it enables tachyon universes to spawn in a new novel way not previously considered in discussions of the origin of universes.

A simple model lagrangian[136] for a tachyon universe particle is

---

[135] We will use the term mass here to denote mass-energy. Since we identified mass as a multiple of area earlier, the comments here would appear to apply to universe area as well.

[136] See Blaha (2018e) and earlier books by the author for a detailed discussion.

$$\mathcal{L}_{\parallel} = \psi_T^{\ S}(Y(y))[\gamma^\mu \partial/\partial y^\mu - e_B \gamma^\mu B_{u\mu}(Y(y)) - m(t)]\psi_T(Y(y)) - \tfrac{1}{4}\, F_{Bu}^{\ \mu\nu}(Y(y))F_{Bu\mu\nu}(Y(y))$$
$$- \tfrac{1}{4}\, F_u^{\ \mu\nu}(y)F_{u\mu\nu}(y)$$

We assume m(t) is constant.

When a particle or a universe particle fissions (decays) one normally expects that the masses of the particles or universe particles produced by the decay to be smaller than the mass of the original particle or nucleus. In the case of tachyon (faster-than-light) elementary particles, or universe particles, a much different possibility is present: a tachyon universe can decay into heavier tachyons (perhaps through a distortion of the universe internally into two 'lumps'.) We consider the specific case of a tachyon universe particle decaying into two universe particles whose total mass is greater than the original. (See Fig. 26.3.)

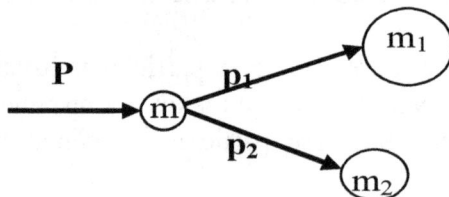

Figure 26.3. Two universe particle decay of a tachyon universe particle.

We will assume the initial tachyon universe particle has zero energy ($p^D = 0$) and thus the tachyons universe particles emerging from the decay also have total universe particle energy zero. The analysis is based on conservation of total universe energy and momentum in Megaverse space outside of universes. The below discussion applies to D-dimensional space with (D – 1)-dimensional spatial coordinates.

Momentum conservation implies

$$\mathbf{P} = \mathbf{p}_1 + \mathbf{p}_2$$

Since all energies are zero

$$(c\mathbf{P})^2 = (c\mathbf{P})^2 = m^2$$

$$(cp_1)^2 = (c\mathbf{p}_1)^2 = m_1^2$$
$$(cp_2)^2 = (c\mathbf{p}_2)^2 = m_2^2$$

where $P = |\mathbf{P}|$, $p_1 = |\mathbf{p}_1|$, and $p_2 = |\mathbf{p}_2|$. If we now square the above equation for $\mathbf{P}$ and then use the above three equations we obtain

$$m^2 = m_1^2 + m_2^2 + 2m_1m_2 \cos \theta$$

where θ is the opening angle between the emerging universe particles momenta of $\mathbf{p}_1$ and $\mathbf{p}_2$. There are a number of interesting cases:

Case $\theta = 0$:
$$m = m_1 + m_2$$

The masses of the outgoing universe particles sum to the mass of the original tachyon universe particle.

Case $\theta = \pi/2$:
$$m^2 = m_1^2 + m_2^2$$

The masses of each outgoing universe particle tachyon are less than the mass of the original tachyon universe particle.

Case $\theta = \pi$:
$$m^2 = (m_1 - m_2)^2$$

In this case either $m_1 > m$ or $m_2 > m$. Thus one of the outgoing tachyon universe particles has a greater mass than the original tachyon universe particle. Mass is effectively created from the spatial momentum of the initial universe particle. This process is the inverse of normal particle and universe particle fission where the sum of the outgoing masses is always less than the original particle's mass and the difference is mass converted into energy in the form of additional photons.

This last case, where one of the outgoing universe particles is more massive than the original universe particle, is not just for $\theta = \pi$. Since

$$\cos\theta = (m^2 - m_1^2 - m_2^2)/(2m_1 m_2)$$

we see that the sum of the outgoing universe particle masses is always greater than the original tachyon universe particle *mass (except when $\theta = 0$)* since

$$\cos\theta = 1 + [m^2 - (m_1 + m_2)^2]/(2m_1 m_2) \le 1$$

and thus

$$[m^2 - (m_1 + m_2)^2]/(2m_1 m_2) \le 0$$

Note $m = m_1 + m_2$ only if $\theta = 0$.

Since we can transform the above discussion to the case of universe particle tachyons having non-zero Megaverse energy using an ordinary D-dimensional Lorentz transformation, the discussion in this subsection is general.

We therefore conclude that when a tachyon universe particle decays into two tachyon universe particles the sum of the masses of the produced tachyon universe particles is greater than the mass of the original tachyon universe particle except if the angle between the momenta of the produced tachyon universe particles is zero. In that case the sum of the masses of the produced tachyon equals the mass of the original tachyon universe particle and the produced universe particles overlap.

Appendix A. Some Physical Constants Used in Calculating Numerical Expressions

Some physical constants that we found to be of use in the evaluation of expressions are (assuming units with $c = \hbar = 1$):

$$G \equiv 7.39 \times 10^{-29} \text{ gm}^{-1} \text{ cm} \tag{A.1a}$$
$$G \equiv 2.6 \times 10^{-66} \text{ cm}^2 \tag{A.1b}$$
$$G \equiv 2.91 \times 10^{-87} \text{ s}^2 \tag{A.1c}$$

$$H_0 \equiv 2.133 \times 10^{-33} h \text{ ev} \tag{A.2a}$$
$$H_0 \equiv 1.08 \times 10^{-28} h \text{ cm}^{-1} \tag{A.2b}$$
$$H_0 \equiv 3.24 \times 10^{-18} h \text{ s}^{-1} \tag{A.2c}$$
$$h = .689 \tag{A.2d}$$
$$H_0 = 100 \ h \text{ km s}^{-1} \text{ Mpc}^{-1} \tag{A.2e}$$
$$1 \text{ s}^{-1} = 3.0864 \times 10^{-20} \text{ km s}^{-1} \text{ Mpc}^{-1} \tag{A.2f}$$

$$GH_0 \equiv 7.98 \times 10^{-57} h \text{ gm}^{-1} \tag{A.3}$$

$$\rho_{\text{crit}} \equiv 1.88 \times 10^{-29} h^2 \text{ gm cm}^{-3} \tag{A.4}$$

$$M_{\text{Planck}} \equiv 1.22 \times 10^{28} \text{ ev} \tag{A.5a}$$
$$M_{\text{Planck}} \equiv 2.18 \times 10^{-5} \text{ g} \tag{A.5b}$$
$$M_{\text{Planck}} \equiv 6.20 \times 10^{32} \text{ cm}^{-1} \tag{A.6a}$$
$$\text{Planck Length} = M_{\text{Planck}}^{-1} \equiv 1.61 \times 10^{-33} \text{ cm} \tag{A.6b}$$
$$M_{\text{Planck}} \equiv 1.85 \times 10^{43} \text{ s}^{-1} \tag{A.7a}$$
$$\text{Planck time} = M_{\text{Planck}}^{-1} \equiv 5.41 \times 10^{-44} \text{ s} \tag{A.7b}$$

$$1 \text{ eV} \equiv 5.08 \times 10^4 \text{ cm}^{-1} \tag{A.8a}$$

$$1 \text{ eV} \equiv 1.52 \times 10^{15} \text{ s}^{-1} \tag{A.8b}$$

$$1 \text{ eV/c}^2 \equiv 1.783 \times 10^{-33} \text{ g} \tag{A.8c}$$

$$1 \text{ g} \equiv 2.85 \times 10^{37} \text{ cm}^{-1} \tag{A.8d}$$

$$\kappa \equiv 4.38 \text{ °K}^{-1} \text{ cm}^{-1} \tag{A.9}$$

$$\kappa \equiv 1.31 \times 10^{11} \text{ °K}^{-1} \text{ s}^{-1} \tag{A.10}$$

$$\kappa \equiv 8.62 \times 10^{-5} \text{ ev °K}^{-1} \tag{A.11}$$

$$1 \text{ Gyr} = 3.16 \times 10^{16} \text{ s} \tag{A.12}$$

where κ is Boltzmann's constant.

Appendix B. Defined Quantities in the Text

This appendix lists constants defined by the Particle Data Group and in the text.

Constants based on M. Tanabashi *et al op. cit.*

h = Hubble Constant = 0.678(9)

ρ_{crit} = Critical density = 1.87840(9) $h^2 \times 10^{-29}$ g/cm^{-3}

Ω_Λ = Dark Energy density/ρ_{cr} = ρ_{de}/ρ_{cr} = 0.692 ± 0.012

$\Omega_d = \Omega_c$ = Cold Dark matter density/ρ_{cr} = ρ_c/ρ_{cr} = 0.1186(20) h^{-2}

Ω_b = Baryon density/ρ_{cr} = ρ_b/ρ_{cr} = 0.02226 h^{-2}

$\Omega_m = \Omega_b + \Omega_c$

$\quad$ = pressureless Matter density/ρ_{cr} = ρ_m/ρ_{cr} =0.308 ± 0.012

$\Omega_{mtot} = \Omega_m + \Omega_\Lambda$ = 1.000 ± 0.024

$t_0 = t_{now}$ = Age of universe = 13.80 ± 0.04 Gyr

$r_{universe}(t_{now})$ = visible radius of the universe = 4.314 $\times 10^{28}$ cm $\qquad$ (13.1.3)

Ω_γ = radiation density/ρ_{cr} = ρ_γ/ρ_{cr} = 2.473$h^{-2} \times 10^{-5}$ $(T/2.7255)^4 h^{-2}$

$\quad$ = 5.38 $\times 10^{-5}$ $\hspace{4cm}$ (13.1.4)

k = 5.56 $\times 10^{-57}$ cm^{-2}

Constants defined in the text:

$$C_M = C_T G \rho_{Mega} \qquad\qquad (9.8)$$

Robertson-Walker volume:
$$V = 2\pi^2 R^3 \qquad\qquad (10.1.2b)$$

$$\alpha = \gamma/\delta = 1 \qquad\qquad (10.6.3)$$

From the current CMB temperature ($T_0 = 2.7255\ ^\circ$K) we find

$$\kappa T = \kappa T_0/a(t_{now}) = \kappa T_0 = 2.35 \times 10^{-4}\ ev \qquad (13.3.1.1)$$
$$\chi = \pi^{3/2}\kappa T_0/(2M_c) \cong 5.3 \times 10^{-32} \qquad (13.3.1.2)$$

$$H_0 = 100\ h\ km\ s^{-1}\ Mpc^{-1} = h \times (9.777752\ Gyr)^{-1} =$$
$$= 1.1 \times 10^{-28}h\ cm^{-1} \equiv 3.24 \times 10^{-20}h\ s^{-1} \qquad (13.2.1.2b)$$
$$= 5.56 \times 10^{-57}\ cm^{-2} \qquad (13.2.1.2c)$$

$$\Omega_m + \Omega_\Lambda \cong 1.00$$

$$\xi = k/H_0^2 \cong 1 \qquad (13.2.1.3)$$

$$a(t_{now}) = 1 \qquad (13.2.2.4)$$

$$H_0\Omega_\Lambda^{\frac{1}{2}} = 1.83 \times 10^{-18}\ s^{-1} \qquad (13.2.2.6)$$

$$t_E = t_{now} + (H_0\Omega_\Lambda^{\frac{1}{2}})^{-1}\ln(.76) = 2.87 \times 10^{17}\ s \qquad (13.2.2.7)$$
$$t_{now} = 4.35 \times 10^{17}\ s$$
$$t_{RM} = 7 \times 10^{11}\ s = 2.22 \times 10^{-5}\ Gyr \qquad 13.2.3.12)$$
$$t_c = 1.26 \times 10^{-165}\ s \cong 10^{-165}\ s \qquad (13.3.2.15)$$
$$t_T = 380,000\ years$$
$$= 1.19837 \times 10^{13}\ sec$$
$$= 1.205 \times 10^{-11}\ Gyr$$

$$\kappa T = \kappa T_0/a(t_{now}) = \kappa T_0 = 2.35 \times 10^{-4}\ ev \qquad (13.3.1.1)$$

$$\varpi = k^{-\frac{1}{2}}M_c = 8.31 \times 10^{60} \qquad (13.3.2.2)$$

$$\gamma = \chi/\varpi = 6.38 \times 10^{-93} \qquad (13.3.2.9)$$

Appendix C. Introduction to Universe Particles

C.1 Megaverse Dynamics due to Universe Fields

The particles and fields that we have found in our universe will also exist in the Megaverse but the form of their Fourier expansions will be 192-dimensional. There will also be universe particles – quantum fields for entire universes.[137]

The universe particles within a Megaverse have dynamical motions in the Megaverse due to forces between them as well as Megaverse gravity.

In Blaha (2017c) as well as earlier books we described features of the Wheeler-DeWitt equation that suggested that universes could be viewed as particles or anti-particles, or tachyons. The solutions of their equations are wave functions on a manifold that are analogous to the solutions of the Klein-Gordon and Dirac equations. The issues of negative probabilities, possible tachyon solutions, and negative frequency solutions suggest a need for an appropriate particle interpretation of universes that can possibly resolve these problems.

Some physicists have taken the Wheeler-DeWitt equation as the starting point for a theory of a universe as a particle. The Wheeler-DeWitt equation describes the interior of a universe in a quantum framework.

We will take a different approach using the Megaverse as the environment of universe particles that internally have Quantum Gravity, and externally have Megaverse Quantum Gravity.

We view a universe as an extended particle and begin by ignoring the detailed inner structure of universes. This approach is similar to the historical treatment of hadrons such as the proton as particles and developing a theory of them as fundamental particles using form factors, structure functions and so on to approximate their inner structure. Afterwards, as detailed data became available, the detailed investigation of the internal structure of hadrons using quark-parton models followed. We will pursue a

---

[137] Discussed later.

similar theoretical development beginning with a theory of universes as extended particles in the 192-dimension Megaverse.[138] The internal structure of the particle universes will eventually be specified by the Wheeler-DeWitt equation expressed in Megaverse coordinates.

The two simplest choices for the nature of universes are "spin 0" *boson universes* and *fermionic universes* with odd half integer spin, s_M.[139] We will first consider the possibility of fermion universes, and then consider "spin 0" *boson universes*.

The first issue of fermion universes (reminiscent of the discussions of spin in the 1920's) is the interpretation of spin states. We suggest that the upper $2^{D/2 - 1}$ components (with $2^{D/2 - 2}$ "spin up" and $2^{D/2 - 2}$ "spin down" states) of a fermion universe wave function represent a left-handed universe with an excess number of baryons. The lower $(2^{D/2 - 1})$ components lead to right-handed anti-universes where there is an excess of anti-baryons. These associations are analogous to the interpretations of the Dirac electron wave function.[140]

The universe particle "spin up" and "spin down" states are distinguished by their interactions with gauge fields in a manner analogous to quantum electrodynamics.

C.2 "Free Field" Dynamics of Fermion Universe Particles

We now consider fermion universes as extended particles with an odd half integer spin – *fermion universe particles* - in the D-dimensional Megaverse. In the Megaverse there are D 'Dirac' matrices with $2^{D/2}$ rows and $2^{D/2}$ columns that are the equivalent of the four Dirac matrices in four dimensions. We will denote these D matrices as γ_M^i for i = 1, 2, … , D. They satisfy the anti-commutation relations:

$$\{\gamma_M^i, \gamma_M^j\} = 2\,\delta^{ij} \qquad (C.1)$$

---

[138] Of course we follow a different course here – looking from within universe particles 'outwards' whereas in our universe we look 'into' particles from 'outside.'

[139] The Megaverse is assumed to be D-dimensional where we have assumed provisionally to be D = 192. The spin of fermionic universe particles was shown to be s_M in Blaha (2017c) in chapter 3.

[140] It is known that phenomena in our universe tend to be left-handed. If this feature of our universe's phenomena is also a property of the universe itself, then, since handedness is an attribute of spin, the treatment of a universe as having spin is not unreasonable.

and thus form a Clifford algebra. We will choose the coordinate y^D to be the time coordinate and thus make it pure imaginary with a Reality group transformation. (The D-dimensional Megaverse space is a complex Euclidean space.) Therefore γ^D will be hermitean $((\gamma^D)^2 = 1)$, and the γ^i matrices for i = 1, ... , (D – 1) will be anti-hermitean with $(\gamma^i)^2 = -1$. The number of linearly independent matrices in D dimensions is 2^D.

The Megaverse metric is (by using the Reality group) chosen to be

$$g^{ij} = -\delta^{ij}, \qquad g^{D,D} = 1 \qquad (C.2)$$

for i, j = 1, 2, ... , (D – 1); and zero otherwise.

Except for the additional dimensions, fermion dynamics is quite similar to the 4-dimensional case. The free universe particle Dirac equation is

$$(i\gamma^i\partial_i + m)\psi(y) = 0 \qquad (C.3)$$

summed over i = 1, 2, ... , D where the mass is assumed to be constant, and set below. The derivative operator, is based on the use of quantum coordinates[141]

$$X^i(y) = y^i + i\,Y_u^{\,i}(y)/M_u^{D/2} \qquad (C.4)$$

for i = 1, ... , D and is defined to be

$$\partial_i = \partial/\partial X^i(y) = \partial/\partial(y_i - Y_{ui}(y)/M_u^{D/2}) \qquad (C.5)$$

where *we assume $M_u = M_c$ with M_c being a very large mass scale of perhaps the order of the Planck mass.*

$Y_u^{\,i}$ is a D-dimensional Megaverse gauge field equivalent of the universe $Y^\mu(x)$ used in 4-dimensional Two-Tier renormalization:

$$X^\mu(z) = z^\mu + i\,Y^\mu(z)/M_c^2 \qquad (C.6)$$

---

[141] Giving Two-Tier renormalization without infinities.

where $Y^\mu(z)$ is a free QED-like field. The $Y^i(y)$ quantum coordinates will be used in the Megaverse to eliminate potential divergences, in a manner similar to the case of our universe when universe particle interactions are introduced later. In a high dimension space the potential divergences in calculations are much greater. *Two-Tier coordinates eliminate these potential divergences.*

C.3 Four Types of Fermion Universe Particles

Assuming universe energies are real-valued,[142] there are four possible types of fermion universe particles in the Megaverse. They are analogous to the four species of fermions. Two of these types are tachyonic. It is important to note that DeWitt points out that the Wheeler-DeWitt equation has tachyon solutions since the mass-like term dependent on $^{(3)}R$ can be positive or negative.[143] A negative mass is an indication of tachyon behavior wherein the wave propagation of the state functional is not necessarily in time-like directions and is thus tachyon.

Eq. C.3 is a D-dimensional Dirac equation. There are three other general types of universe particle equations. (By assumption fermion universes come in four species like fermions.) The derivation of the four types of universe particles (appendix F) is similar to the derivation of fermion types discussed previously. We will now consider the D-dimensional equivalent for universe particles in the Megaverse.

The general form of a pure D-dimensional complex Lorentz group[144] boost can be expressed in terms of a complex relative $(D-1)$-velocity $\mathbf{v_c}$ between inertial reference frames. A D-dimensional coordinate boost has the form

$$\Lambda_C(\mathbf{v_c}) \equiv \Lambda_C(\omega, \mathbf{v_c}) = \exp[i\omega\hat{\mathbf{w}}\cdot\mathbf{K}] \tag{C.7}$$

---

[142] The energy of universe particles need not be real-valued since universes can 'decay' – unlike elementary particles which are not subject to decay, by definition, since they are assumed to be *fundamental*. We choose to consider the case of universes with real-valued energies. The case of universes with complex-valued energies is a simple extension of the real-value cases considered here.

[143] DeWitt, B. S., Phys. Rev. **160**, 1113 (1967) p. 1124.

[144] The D-dimensional Complex Lorentz group has similar features to the 4-dimensional Complex Lorentz group. We shall only discuss it to the extent needed for our universe particle type's derivation. See Weinberg (1995) for the 4-dimensional Lorentz group – the D-dimensional Lorentz group generalizes directly from the features of the 4-dimensional Lorentz group.

where

$$\omega = (\omega_r{}^2 - \omega_i{}^2 + 2i\omega_r\omega_i \ \hat{\mathbf{u}}_r{\cdot}\hat{\mathbf{u}}_i)^{\frac{1}{2}} \qquad (C.8)$$

and

$$\hat{\mathbf{w}} = (\omega_r\hat{\mathbf{u}}_r + i\omega_i\hat{\mathbf{u}}_i)/\omega \qquad (C.9)$$

with all vectors being $(D - 1)$-dimensional spatial vectors. We define the real and imaginary unit vectors $\hat{\mathbf{u}}_r{\cdot}\hat{\mathbf{u}}_r = 1 = \hat{\mathbf{u}}_i{\cdot}\hat{\mathbf{u}}_i$ with the result

$$\hat{\mathbf{w}}{\cdot}\hat{\mathbf{w}} = 1 \qquad (C.10)$$

The complex relative velocity is

$$\mathbf{v}_c = \hat{\mathbf{w}} \ \tanh(\omega) \qquad (C.11)$$

The free dynamical equations of the four universe particle species will be generated by D-dimensional Lorentz boosts of the free Dirac equation of a universe particle at rest with the *requirement that the time variable* $(t = y^D)$ *and energy are real in the resulting field equations.*[145] The procedure can most easily be performed in D-dimensional momentum space with the Megaverse coordinate space version of the generated equation determined from the momentum space version.

The result is a number of different types of fermion universes that mirror the particle species that we have seen earlier.

- Dirac-like Universe Particle
- Tachyon Universe particle
- Left-Handed Tachyon Universe Particles
- Right-Handed Tachyon Universe Particles
- "Up-Quark-like" Universe Particles
- Left-Handed "Down-Quark-like" Tachyon Universe Particles
- Right-Handed Down-Quark-like Tachyon Universe Particles

---

[145] The D-dimensional "energy" must be real since it relates to the area of the universe – a real number.

Free lagrangians, which are similar to the corresponding fermion lagrangians seen earlier, follow for these universe particles.

C.4 Embedding Universes Within the Megaverse

Earlier we developed a view of the general nature of universes in the Megaverse. We now examine the general nature of the Megaverse itself. Consistency *suggests that complex-valued coordinates universes, such as our universe, require an embedding in a complex-valued Megaverse* space. In chapter 31 we determined the provisional dimension of the Megaverse, D = 192, from the dimensions implied by interactions within our universe, and from a need to extend Ω-group symmetry to Megaverse coordinates.

We now assume the Megaverse has an analytic, complex, symmetric metric g_{ik} which satisfies

$$g_{ij} = g_{ji} \tag{C.12}$$

for i, j = 1, 2, ... , D.

The embedding equations for our curved complex-valued universe {x} within the complex, D-dimensional Megaverse {y} are:

$$z_i = f_i(x) \tag{C.13}$$

where x^μ are the complex-valued 4-dimensional coordinates of a universe.
The infinitesimal Megaverse invariant distance is

$$ds^2 = g_{ik}dz^i dz^k \tag{C.14}$$

where the z_i are the complex-valued D-dimensional coordinates of the Megaverse. g_{ik} is the complex-valued Megaverse metric tensor, ds is the D-dimensional invariant Megaverse distance,[146] and the complex-valued, 4-dimensional metric of a universe has the form

---

[146] We note that ds^2 is complex and this might be found troubling. A Reality group transformation would yield the absolute value of ds^2 and thus $ds_{physical} = \|ds^2\|^{\frac{1}{2}}$ would be the physical real-valued invariant distance. This remark also applies to 4-dimensional complex General Relativity.

$$g_{\mu\nu} = \partial f_j/\partial x^\mu \; \partial f_j/\partial x^\nu \qquad\qquad (C.15)$$

with an implied sum over the subscript j.

The picture that we paint of the Megaverse is that of a complex, D-dimensional space containing (perhaps) countless universes, some of which may be (almost) flat like ours, and some of which may be curved, open or closed 4-surfaces. *We assume universes are closed, or curved and open, or flat.*

The Megaverse does have a gravitational field due to universe masses, particle masses, the Two-Tier[147] field Y^μ, and the Unified SuperStandard Theory fields. *We will assume all universes are complex, 4-dimensional although universes with a larger number of dimensions are possible and will be discussed below.*

C.5 Possible Dimensions of Universes

In Blaha (2011c) (and in this book) we determined the dimensionality of our universe based on principles of Asynchronous Logic that suggested a 4-valued logic that could be embodied in a 4-dimensional spinor matrix formulation. This 4-dimensional spinor formulation led to a 4-dimensional space-time.

The requirement that the speed of light is the same in all inertial reference frames, and that transformations between reference frames in faster than light relative motion are physically allowed, led to the requirement of complex coordinates and the Complex Lorentz transformation group (as was found necessary in Axiomatic Quantum Field Theory studies.) The reality of all physical time and distance measurements led to the introduction of the Reality group that mapped complex quantities to real physical values. This chain of logic is in accord with Leibniz's minimax principle: nature uses the simplest means to create complex physical phenomena.

While the preceding paragraph applies very nicely to our universe, and presumably to other universes, the question of the existence of other universes within the Megaverse with different dimensions naturally arises. Stars and galaxies have many varieties. Why should all universes be of dimension four?

Having developed the fundamental nature of our universe from Logic (the only sure requirement of any physical theory) it seems reasonable to classify possible universes based on their fundamental logic. In Blaha (2011c) we developed matrix formulations for many-valued logics. Assuming no separate clock mechanism to synchronize parts of complex processes, we developed an n × n matrix formalism for n-valued logic.

Therefore we were able to develop a principal sequence of types of universes based on n-valued logic. We summarize the small n-valued cases in Table C.1.

Fant (2005) points out that VLSI circuits with spatially separated parts, which require time synchronization of activity without clocks, need a 4-valued logic at minimum. Thus for a complex universe such as ours, the minimum space-time dimensionality is 4. For a smaller number of dimensions the complexity of physical processes is much diminished as the many solvable models of low space-time dimensionality show: easily solved – not very complex phenomena!

Smaller dimensioned universes may well exist – but not with the richness of complexity that leads to our type of universe's phenomena such as life.

Larger dimension universes may well also exist. They would have an excess of phenomena that might preclude life as we know it, or engender new forms of life.

The general tendency of physical phenomena to be largely based on extrema suggests that 4-dimensional space-time based on Leibniz's minimax principle is the "logical" choice for all universes.The case of 6-dimensional universes also appears attractive for a number of reasons.

The above classification scheme for universes is based on logic. Another important consideration is size. It appears that universes can have differing sizes and in fact can also grow or diminish in size (expansion or contraction). We have considered the size issue earlier and will again when we consider the expansion/contraction of a universe due to a time-dependent mass. Other possible differentiating factors between universes will also be considered.

n-Valued Logic	Matrix Representation Size	Spinor Components	Space-Time Dimensionality[148]
1	1×1	1	1
2	2×2	2	2
3	3×3	2	3
4	4×4	4	4
5	5×5	4	5
6	6×6	8	6

Table C.1. Space-time dimensionality and number of spinor components corresponding to various n-valued logics. It would seem that the minimal acceptable number of dimension is 4 if one is to have a physically acceptable universe as we understand it.

C.6 Fermion Spins in the Megaverse for D = 192

The spinors in free field expansions of $\psi(y)$ have $2^{D/2}$ column entries.[149] The D 'Dirac' γ-matrices have $2^{D/2} \times 2^{D/2}$ rows and columns. The lowest fermion spin is

$$s_M = (2^{D/2 - 1} - 1)/2 \tag{C.16}$$

For D = 192 these quantities are (rounded off – they are half-integer values in fact):

Number of spinor components:
$N_{MRC} = 2^{D/2} = 8 \times 10^{28}$ (C.17)
Spin:
$s_M = 4 \times 10^{28}$
Spin Range
$s_{Mz} = -4 \times 10^{28}, -4 \times 10^{28} + 1, \ldots, +4 \times 10^{28}$

---

[148] Weinberg (1995) p. 216 exhibits an equation that relates the number of components of a spinor to the dimension of its space-time.
[149] See Weinberg (1995) p. 216.

At first glance the range of spins gives one pause for thought. However, universes are not small generally. So a large range of spins is understandable.

Appendix D. Fermion and Boson Universe Particles

The Wheeler-DeWitt equation, because of its similarity to the Klein-Gordon equation, has led to numerous proposals to view universes as particles.[150]

In this chapter[151] we will consider a possible particle interpretation of universes that, while consistent with the spirit of the Wheeler-DeWitt equation and the Megaverse, goes far beyond our current experimental knowledge, although some recent astronomical data tends to support it. It can only be justified in this century by its generality and simplicity. It just looks right.

D.1 The Hierarchy of the Cosmos

In our universe we have seen that natural phenomena form a hierarchy ranging from the simplest to the largest/most complex phenomena. One current view of the hierarchy of levels of physical phenomena is:

Elementary particles: leptons, quarks, gluons, gauge bosons, and Higgs particles
Hadrons: protons, neutrons, …
Molecules
Agglomerations of molecules
Macroscopic objects
Planets
Stars
Galaxies

---

[150] Some suggestions of this interpretation are: DeWitt, B. S., Phys. Rev. **160**, 1113 (1967); Robles-Perez, S. J., arXiv:1212.4598 (2012); and references therein.

[151] This chapter is obtained from Blaha (2015a) with a change of dimension from 16 to D.

Clusters of galaxies
Superclusters
The Universe

Each level generally has a set of "simplified" physical laws that describe its phenomena.[152] For example molecules have quantum mechanical laws and regularities that help to understand the phenomena at the molecular level.

Interestingly, while all phenomena at each level should be explainable by the laws at lower levels, and ultimately, all phenomena should be explainable at the level of elementary particles, connecting phenomena at different levels is often quite difficult and, in many cases, impossible.

Consequently, while we believe physical phenomena are ultimately reducible to the lowest level, the problem of relating phenomena at different levels is largely unresolved.

In this book we introduce new levels in the hierarchy of nature: the level of multiple universes, and the level of the all-encompassing Megaverse. In doing this, we seek to maintain what we know of our universe, as embodied in our Extended Standard Model and Quantum Gravity. We will now turn to a discussion of the universes level and a portrayal of universes as extended particles.

D.2 The Particle Interpretation of Extended Wheeler-DeWitt Equation Solutions

In earlier chapters we described features of the Wheeler-DeWitt equation that suggested that universes could be viewed as particles or anti-particles, or tachyons. The solutions of this equation are scalar wave functions on a manifold that are analogous to the solutions of the Klein-Gordon equation. The issues of negative probabilities, possible tachyon solutions, and negative frequency solutions suggest a need for an appropriate particle interpretation of universes that can possibly resolve these problems.

---

[152] This point was often made by Nobelist Kenneth Wilson of Cornell and Ohio State Universities.

Some physicists have taken the Wheeler-DeWitt equation as the starting point for a theory of a universe as a particle. The Wheeler-DeWitt equation describes the interior of a universe in a quantum framework.

We will take a different approach using the Megaverse as the environment of universe particles that internally have Quantum Gravity, and externally have Megaverse Quantum Gravity.

We view a universe as an extended particle and begin by ignoring the detailed inner structure of universes. This approach is similar to the historical treatment of hadrons such as the proton as particles and developing a theory of them as fundamental particles using form factors, structure functions and so on to approximate their inner structure. Afterwards, as detailed data became available, the detailed investigation of the internal structure of hadrons using quark-parton models followed. We will pursue a similar theoretical development beginning with a theory of universes as extended particles in the D-dimension Megaverse. The internal structure of the particle universes will eventually be specified by the Wheeler-DeWitt equation expressed in Megaverse coordinates.

The two simplest choices for the nature of universes are "spin 0" *boson universes* and *fermion universes* with odd half integer spin, s_M.[153] We will first consider the possibility of fermion universes, and then briefly consider "spin 0" *boson* universes.

The first issue of fermion universes (reminiscent of the discussions of spin in the 1920's) is the interpretation of spin states. We suggest that the upper $2^{D/2-1}$ components (with $2^{D/2-2}$ "spin up" and $2^{D/2-2}$ "spin down" states) of a fermion universe wave function represent a left-handed universe with an excess number of baryons. The lower $(2^{D/2-1})$ components lead to right-handed anti-universes where there is an excess of anti-baryons. These associations are analogous to the interpretations of the Dirac electron wave function.[154]

The universe particle "spin up" and "spin down" states are distinguished by their interactions with gauge fields in a manner analogous to quantum electrodynamics.

---

[153] Since the Megaverse is D-dimensional, the spin of fermionic universe particles was shown to be s_M in chapter 37.

[154] It is known that phenomena in our universe tend to be left-handed. If this feature of our universe's phenomena is also a property of the universe itself, then, since handedness is an attribute of spin, the treatment of a universe as having spin is not unreasonable.

D.3 "Free Field" Dynamics of Fermion Universe Particles

We now consider universes as extended particles with an odd half integer spin – *fermion universe particles* - in the D-dimensional Megaverse. In the Megaverse there are D 'Dirac' matrices with $2^{D/2}$ rows and $2^{D/2}$ columns that are the equivalent of the four Dirac matrices in four dimensions. We will denote these D matrices as γ_M^i for i = 1, 2, … , D. They satisfy the anti-commutation relations:

$$\{\gamma_M^i, \gamma_M^j\} = 2\,\delta^{ij} \tag{D.1}$$

and thus form a Clifford algebra. We will choose y^D to be the time coordinate and thus make it pure imaginary with a Reality group transformation. (The D-dimensional Megaverse space is a complex Euclidean space.) Therefore γ^D will be hermitean $((\gamma^D)^2 = 1)$, and the γ^i matrices for i = 1, … , (D – 1) will be anti-hermitean with $(\gamma^i)^2 = -1$. The number of linearly independent matrices in D dimensions is 2^D.

The Megaverse metric is (by use of the Reality group) chosen to be

$$g^{ij} = -\delta^{ij}, \qquad g^{D,D} = 1 \tag{D.2}$$

for i, j = 1, 2, … , (D – 1); and zero otherwise.

Except for the additional dimensions, fermion dynamics is quite similar to the 4-dimensional case. The free universe particle Dirac equation is

$$(i\gamma^i\partial_i + m)\psi(y) = 0 \tag{D.3}$$

summed over i = 1, 2, … , D where the mass is assumed to be constant, and set by eq. D.119 below. The derivative operator, is based on the use of quantum coordinates[155]

$$Y^i(y) = y^i + i\,Y_u^i(y)/M_u^{D/2} \tag{D.4}$$

For i = 1, …, D and is defined to be

---

[155] Giving Two-Tier renormalization.

$$\partial_i = \partial/\partial Y^i(y) = \partial/\partial(y_i - Y_{ui}(y)/M_u^{D/2}) \qquad (D.5)$$

where *we assume $M_u = M_c$ with M_c being a very large mass scale of perhaps the order of the Planck mass.*

Y_u^i is a D-dimensional Megaverse gauge field equivalent of the universe $Y^\mu(x)$ used in 4-dimensional Two-Tier renormalization (discussed in I):

$$Y^\mu(y) = y^\mu + i\, Y^\mu(y)/M_c^2$$

where $Y^\mu(y)$ is a free QED-like field. The $Y^i(y)$ quantum coordinates will be used in the Megaverse to eliminate potential divergences, in a manner similar to the case of our universe when universe particle interactions are introduced later.

D.3.1 Four Types of Fermion Universe Particles

Assuming universe energies are real-valued,[156] there are four possible types of fermion universe particles in the Megaverse that are analogous to the four species of fermions described in Blaha (2010b) for The Extended Standard Model. Two of these types are tachyonic. It is important to note that DeWitt points out that the Wheeler-DeWitt equation has tachyon solutions since the mass-like term dependent on $^{(3)}R$ can be positive or negative.[157] A negative mass is an indication of tachyon behavior wherein the wave propagation of the state functional is not necessarily in time-like directions and is thus tachyon.

Eq. D.3 is a Dirac-type D-dimensional Dirac equation. There are three other general types of universe particle equations. (By assumption fermion universes come in four species like fermions.) The derivation of the four types of universe particles is similar to the derivation of fermion types in the Extended Standard Model in 4-

---

[156] The energy of universe particles need not be real-valued since universes can 'decay' – unlike elementary particles which are not subject to decay, by definition, since they are assumed to be *fundamental*. We choose to consider the case of universes with real-valued energies. The case of universes with complex-valued energies is a simple extension of the real-value cases considered here.

[157] DeWitt, B. S., Phys. Rev. **160**, 1113 (1967) p. 1124.

dimensional complex space-time given in Blaha (2010b). We will now consider the D-dimensional equivalent for universe particles in the Megaverse.

The general form of a pure D-dimensional complex Lorentz group[158] boost can be expressed in terms of a complex relative (D − 1)-velocity $\mathbf{v_c}$ between inertial reference frames. A D-dimensional coordinate boost has the form

$$\Lambda_C(\mathbf{v_c}) \equiv \Lambda_C(\omega, \mathbf{v_c}) = \exp[i\omega\hat{\mathbf{w}}\cdot\mathbf{K}] \tag{D.6}$$

where

$$\omega = (\omega_r^2 - \omega_i^2 + 2i\omega_r\omega_i\,\hat{\mathbf{u}}_r\cdot\hat{\mathbf{u}}_i)^{\frac{1}{2}} \tag{D.7}$$

and

$$\hat{\mathbf{w}} = (\omega_r\hat{\mathbf{u}}_r + i\omega_i\hat{\mathbf{u}}_i)/\omega \tag{D.8}$$

with all vectors being (D − 1)-dimensional spatial vectors. We define the real and imaginary unit vectors $\hat{\mathbf{u}}_r\cdot\hat{\mathbf{u}}_r = 1 = \hat{\mathbf{u}}_i\cdot\hat{\mathbf{u}}_i$ with the result

$$\hat{\mathbf{w}}\cdot\hat{\mathbf{w}} = 1 \tag{D.9}$$

The complex relative velocity is

$$\mathbf{v_c} = \hat{\mathbf{w}}\tanh(\omega) \tag{D.10}$$

The free dynamical equations of the four universe particle species will be generated by D-dimensional Lorentz boosts of the free Dirac equation of a universe particle at rest with the *requirement that the time variable* (t = y^D) *and energy are real in the resulting field equations*.[159] The procedure can most easily be performed in D-dimensional momentum space with the Megaverse coordinate space version of the generated equation determined from the momentum space version.

---

[158] The D-dimensional complex Lorentz group has similar features to the 4-dimensional complex Lorentz group. We shall only discuss it to the extent needed for our universe particle type's derivation. See Weinberg (1995) for the 4-dimensional Lorentz group – the D-dimensional Lorentz group generalizes directly from the features of the 4-dimensional Lorentz group.

[159] The D-dimensional "energy" must be real since it relates to the area of the universe – a real number.

D.3.1.1 Dirac-like Equation – Type I universe Particle

A positive energy plane wave solution of the Dirac equation eq. D.3 for a universe particle at rest is

$$\psi(y) = \exp[-imt]w(0) \tag{D.11}$$

where we set $\partial_t = \partial/\partial y_D$ while temporarily ignoring the $Y_u^i(y)/M_u^{D/2}$ term. $w(0)$ is a $2^{D/2}$ component spinor column vector. The solution $\psi(y)$ satisfies the momentum space Dirac equation for a particle at rest:

$$(m\gamma^D - m)\psi(y) = 0 \tag{D.12}$$

The $2^{D/2} \times 2^{D/2}$ spinor matrix form of a D-dimensional Lorentz boost with a relative real velocity $\mathbf{v}$ of the Dirac matrices is[160]

$$S^{-1}(\Lambda(\mathbf{v}))\gamma^\nu S(\Lambda(\mathbf{v})) = \Lambda^\nu_\mu(\mathbf{v})\gamma^\mu \tag{D.13}$$

where $\Lambda^\nu_\mu(\mathbf{v})$ is a D-dimensional Lorentz boost. $S(\Lambda(\mathbf{v}))$ has the form

$$S(\Lambda(\mathbf{v})) = \exp(-\omega\gamma^D\boldsymbol{\gamma}\cdot\mathbf{v}/(2|\mathbf{v}|))$$

$$= \cosh(\omega/2)I + \sinh(\omega/2)\gamma^D\boldsymbol{\gamma}\cdot\mathbf{p}/|\mathbf{p}| \tag{D.14}$$

with *real* $\omega = \operatorname{arctanh}(|\mathbf{v}|)$ and *real* $\mathbf{v}$. $|\mathbf{p}|$ is the magnitude of the spatial (D − 1)-vector. Also

$$S^{-1}(\Lambda(\mathbf{v})) = \gamma^D S^\dagger(\Lambda(\mathbf{v}))\gamma^D = \exp(\omega\gamma^D\boldsymbol{\gamma}\cdot\mathbf{v}/(2|\mathbf{v}|))$$

$$= \cosh(\omega/2)I - \sinh(\omega/2)\gamma^D\boldsymbol{\gamma}\cdot\mathbf{p}/|\mathbf{p}| \tag{D.15}$$

---

[160] **The indices ν and μ from this point in this chapter have values: 1, 2, … , D.**

If we now apply $S(\Lambda(\mathbf{v}))$ to the momentum space Dirac equation of a particle at rest (eq. D.12) we find

$$0 = S(\Lambda(\mathbf{v}))(m\gamma^D - m)\,\psi(y)$$
$$= [mS(\Lambda(\mathbf{v}))\gamma^D S^{-1}(\Lambda(\mathbf{v})) - m]S(\Lambda(\mathbf{v}))w(0)$$

A straightforward evaluation shows

$$mS(\Lambda(v))\gamma^D S^{-1}(\Lambda(v)) = g_{D\mu\nu}p^\mu\gamma^\nu = \not{p} \qquad (D.16)$$

where p is a momentum D-vector. In addition we define the D-dimension spinor ($2^{D/2}$ components)

$$S(\Lambda(v))w(0) = w(p) \qquad (D.17)$$

which can be viewed as a "positive energy D Dirac spinor". The Dirac equation in momentum space has the familiar form:

$$(\not{p} - m)\exp[-ip\cdot y]w(p) = 0 \qquad (D.18)$$

Eq. D.18 implies the free, coordinate space Dirac equation:

$$(i\gamma^\mu \partial/\partial y^\mu - m)\psi(y) = 0 \qquad (D.19)$$

We identify this equation as the dynamical equation of a type 1 universe particle. It corresponds to the free charged lepton elementary particle species Dirac equation in particle physics.

D.3.1.2 Complex Boosts

The form of the D-dimensional spinor boost transformation corresponding to the coordinate transformation eq. D.6 is:

$$S_C(\omega, \mathbf{v_c}) \equiv S_C = \exp(-\omega\gamma^D\gamma\cdot\hat{\mathbf{w}}/2)$$

$$= \cosh(\omega/2)I + \sinh(\omega/2)\gamma^D\boldsymbol{\gamma}\cdot\hat{\mathbf{w}} \tag{D.20}$$

with *complex* $\mathbf{v_c}$ and $\hat{\mathbf{w}}$ defined by eqs. D.10 and D.8 respectively. The inverse transformation is

$$S_C^{-1}(\omega, \mathbf{v_c}) = \exp(\omega\gamma^D\boldsymbol{\gamma}\cdot\hat{\mathbf{w}}/2)$$

$$= \cosh(\omega/2)I - \sinh(\omega/2)\gamma^D\boldsymbol{\gamma}\cdot\hat{\mathbf{w}} \tag{D.21}$$

Note that S_C is not unitary just as in the 4-dimensional case.

We now apply a spinor boost to the Dirac equation for a particle at rest in this more general case of complex ω and $\hat{\mathbf{w}}$.

$$0 = S_C(\omega, \mathbf{v_c}))(m\gamma^D – m) \exp[-imt]w(0)$$
$$= [mS_C\gamma^D S_C^{-1} – m] \exp[-imt]S_C w(0) \tag{D.22}$$

where $S_C = S_C(\omega, \mathbf{v_c})$. After some algebra we find

$$mS_C\gamma^D S_C^{-1} = m[\cosh(\omega)\gamma^D – \sinh(\omega)\boldsymbol{\gamma}\cdot\hat{\mathbf{w}}] \tag{D.23}$$

We will use these *complex* boosts to generate the other species' Dirac-like equations.

D.3.1.3 Tachyon Universe particle Dirac Equation

The development of the complex spinor boost transformation (subsection D.3.1.2 above) leads to two possible forms of the tachyon Dirac-like equation. One form will lead to a lagrangian dynamics for left-handed universe particles. The other form leads to a lagrangian dynamics for right-handed universe particles.

D.3.1.4 Type IIa Case: Left-Handed Tachyon Universe Particles

If the real and imaginary relative vectors parts of $\hat{\mathbf{w}}$, namely $\hat{\mathbf{u}}_r$ and $\hat{\mathbf{u}}_i$, are parallel, then $\hat{\mathbf{u}}_r\cdot\hat{\mathbf{u}}_i = 1$ and
$$\omega = \omega_r + i\omega_i \tag{D.24}$$

Eqs. D.23 and D.24 then imply

$$mS_C\gamma^D S_C^{-1} = m[\cosh(\omega_r)\cos(\omega_i) + i\sinh(\omega_r)\sin(\omega_i)]\gamma^D -$$
$$- m[\sinh(\omega_r)\cos(\omega_i) + i\cosh(\omega_r)\sin(\omega_i)]\gamma\cdot\hat{\mathbf{u}}_r \qquad \text{D.25})$$

or

$$mS_C\gamma^D S_C^{-1} = \cos(\omega_i)\gamma\cdot p_r + i\sin(\omega_i)\gamma\cdot p_i \qquad \text{(D.26)}$$

where

$$p_r^0 = m \cosh(\omega_r) \qquad p_i^0 = m \sinh(\omega_r) \qquad \text{(D.27)}$$

and

$$\mathbf{p}_r = m\hat{\mathbf{u}}_r \sinh(\omega_r) \qquad \mathbf{p}_i = m\hat{\mathbf{u}}_r \cosh(\omega_r) \qquad \text{(D.28)}$$

If $\omega_i = 0$, then we recover the momentum space Dirac-like equation. If $\omega_i = \pi/2$, then we obtain the left-handed momentum space tachyon equation:

$$mS_C\gamma^D S_C^{-1} = i\gamma\cdot p_i \qquad \text{(D.29)}$$

and the tachyon energy and momentum expressions

$$\mathbf{p} = m\mathbf{v}\gamma_s \qquad E = m\gamma_s \qquad \text{(D.30)}$$

where $\sinh(\omega) = \gamma_s = (\beta^2 - 1)^{-\frac{1}{2}}$ with $\beta = v/c > 1$. v is the absolute value of the $(D - 1)$ component spatial velocity. Also

$$S_C w(0) = w_C(p) \qquad \text{(D.31)}$$

is a tachyon spinor.

The momentum space tachyon Dirac-like equation is

$$(i\not{p} - m)\exp[-ip\cdot y]w_T(p) = 0 \qquad \text{(D.32)}$$

where $p\cdot y = p^D y^D - \mathbf{p}\cdot\mathbf{y}$ after performing a corresponding boost in the exponential factor. If we apply $i\not{p}$ to eq. D.32 we find the tachyon mass condition is satisfied

$$- E^2 + \mathbf{p}^2 = m^2 \tag{D.33}$$

Transforming back to coordinate space we obtain the "left-handed" *tachyon Dirac-like equation*:

$$(\gamma^\mu \partial/\partial y^\mu - m)\psi_T(y) = 0 \tag{D.34}$$

D.3.1.5 Type IIb Case: Right-Handed Tachyon Universe Particles

If the real and imaginary relative vectors parts of $\hat{\mathbf{w}}$, $\hat{\mathbf{u}}_r$ and $\hat{\mathbf{u}}_i$, are anti-parallel $\hat{\mathbf{u}}_r = -\hat{\mathbf{u}}_i$, then $\hat{\mathbf{u}}_r \cdot \hat{\mathbf{u}}_i = -1$ and

$$\omega = \omega_r - i\omega_i \tag{D.35}$$

then

$$mS_C\gamma^D S_C^{-1} = m[\cosh(\omega_r)\cos(\omega_i) - i\sinh(\omega_r)\sin(\omega_i)]\gamma^D -$$
$$- m[\sinh(\omega_r)\cos(\omega_i) - i\cosh(\omega_r)\sin(\omega_i)]\gamma\cdot\hat{\mathbf{u}}_r \tag{D.36}$$

or

$$mS_C\gamma^D S_C^{-1} = \cos(\omega_i)\gamma\cdot p_r - i\sin(\omega_i)\gamma\cdot p_i \tag{D.37}$$

where

$$p_r{}^D = m\cosh(\omega_r) \qquad p_i{}^D = m\sinh(\omega_r) \tag{D.38}$$

and

$$\mathbf{p}_r = m\hat{\mathbf{u}}_r \sinh(\omega_r) \qquad \mathbf{p}_i = m\hat{\mathbf{u}}_r \cosh(\omega_r) \tag{D.39}$$

If $\omega_i = \pi/2$, then we obtain the right-handed momentum space tachyon equation.[161]

$$(-\gamma^\mu \partial/\partial y^\mu - m)\psi_T(y) = 0 \tag{D.40}$$

---

[161] We note that $\gamma_s = (\beta^2 - 1)^{-\frac{1}{2}}$, *if expressed in terms of ω, has a branch cut extending from $<-\infty, +\infty>$ in the complex ω plane. Thus values of ω with positive imaginary parts are physically different from values of ω with negative imaginary parts.*

D.3.1.6 Type III Case: "Up-Quark-like" Universe Particles

There are two other cases where we can obtain fermion dynamical equations with a *real* time variable and real energy.[162] In one case we set $\hat{\mathbf{u}}_r \cdot \hat{\mathbf{u}}_i = 0$ and have a real ω.

If the real and imaginary relative vectors parts of $\hat{\mathbf{w}}$, namely $\hat{\mathbf{u}}_r$ and $\hat{\mathbf{u}}_i$, are perpendicular, $\hat{\mathbf{u}}_r \cdot \hat{\mathbf{u}}_i = 0$, then

$$\omega = (\omega_r^2 - \omega_i^2)^{\frac{1}{2}} \tag{D.41}$$

Thus ω is either pure real ($\omega_r \geq \omega_i$) or pure imaginary ($\omega_r < \omega_i$).

The momentum space equation generated by the corresponding spinor boost is

$$\{m\cosh(\omega)\gamma^D - m\sinh(\omega)\gamma \cdot (\omega_r\hat{\mathbf{u}}_r + i\omega_i\hat{\mathbf{u}}_i)/\omega - m\}\exp[-imt]w_c(p) = 0 \tag{D.42}$$

Defining the momentum 4-vector

$$p = (p^D, \mathbf{p}) \tag{D.43}$$

where

$$p^D = m\cosh(\omega) \qquad \mathbf{p} = \mathbf{p}_r + i\mathbf{p}_i \tag{D.44}$$

with

$$\mathbf{p}_r = m\omega_r\hat{\mathbf{u}}_r\sinh(\omega)/\omega \quad \mathbf{p}_i = m\omega_i\hat{\mathbf{u}}_i\sinh(\omega)/\omega \tag{D.45}$$
$$\mathbf{p}_r \cdot \mathbf{p}_i = 0 \tag{D.46}$$

then we obtain a positive energy Dirac-like equation

$$[p\cdot\gamma - m]\exp[-imt]w_c(p) = 0$$

---

[162] The requirement of a real energy for a universe is not strict. For a fundamental free particle the energy must be real or the particle would be subject to decay – contrary to its assumed fundamental nature. Universes can 'decay' to 'smaller' universes. Therefore the requirement for real energy can be violated. Nevertheless the requirement for real energy is appealing since it leads to four species of universes strengthening the analogy of universes to elementary particles.

or

$$[p^D\gamma^D - (\mathbf{p_r} + i\mathbf{p_i})\cdot\boldsymbol{\gamma} - m]\exp[-i p\cdot y]w_c(p) = 0 \qquad (D.47)$$

with a complex 3-momentum $\mathbf{p}$ and the 4-momentum mass shell condition:

$$p^2 = (p^D)^2 - \mathbf{p_r}\cdot\mathbf{p_r} + \mathbf{p_i}\cdot\mathbf{p_i} = m^2 \qquad (D.48)$$

Note

$$|\mathbf{v_c}| = |\mathbf{p}|/p^D = [(\mathbf{p_r} + i\mathbf{p_i})\cdot(\mathbf{p_r} + i\mathbf{p_i})]^{\frac{1}{2}}/p^D = \tanh(\omega) \qquad (D.49)$$

and so the Lorentz factor is

$$\gamma = \cosh(\omega) \qquad (D.50)$$

Eq. D.47 is the momentum space equivalent of the wave equation[163]

$$[i\gamma^D\partial/\partial t + i\boldsymbol{\gamma}\cdot(\nabla_r + i\nabla_i) - m]\psi_u(t, \mathbf{y_r}, \mathbf{y_i}) = 0 \qquad (D.51)$$

where $\mathbf{y} = \mathbf{y_r} - i\mathbf{y_i}$, and where the grad operators ∇_r and ∇_i are with respect to $\mathbf{y_r}$ and $\mathbf{y_i}$ respectively. Since $\hat{\mathbf{u}}_r\cdot\hat{\mathbf{u}}_i = 0$ we see that there is a subsidiary condition on the wave function

$$\nabla_r\cdot\nabla_i \, \psi_u(t, \mathbf{y_r}, \mathbf{y_i}) = 0 \qquad (D.52)$$

We note eq. D.52 can be put into covariant form as the difference of two vectors squared (which is a real D-dimensional Lorentz group invariant):

$$[\gamma^D\partial/\partial t + i\boldsymbol{\gamma}\cdot(\nabla_r + i\nabla_i)]^2 - [\gamma^D\partial/\partial t + i\boldsymbol{\gamma}\cdot(\nabla_r - i\nabla_i)]^2 = 4\nabla_r\cdot\nabla_i.$$

We identify eq. D.51 as the dynamical equation of an "up-quark-like" universe particle.

---

[163] The gradient operators ∇_r and ∇_i are 191-dimensional spatial gradient operators.

D.3.1.7 Type IVa Case: Left-Handed "Down-Quark-like" Tachyon Universe Particles
   In this case we set $\hat{\mathbf{u}}_r \cdot \hat{\mathbf{u}}_i = 0$. Then by eq. D.7

$$\omega = (\omega_r^2 - \omega_i^2)^{\frac{1}{2}}$$

Thus ω again starts out either pure real (if $\omega_r \geq \omega_i$) or pure imaginary (if $\omega_r < \omega_i$). In this case we also choose ω real, and then change ω to

$$\omega = (\omega_r^2 - \omega_i^2)^{\frac{1}{2}} \rightarrow \omega' = (\omega_r^2 - \omega_i^2)^{\frac{1}{2}} + i\pi/2 = \omega + i\pi/2$$

by adding $i\pi/2$ to ω since ω is a free parameter. We then proceed as we did in the prior tachyon case.[164]. The resulting Lorentz boost

$$\Lambda_C = \exp[i((\omega_r^2 - \omega_i^2)^{\frac{1}{2}} + i\pi/2)(\omega_r \hat{\mathbf{u}}_r + i\omega_i \hat{\mathbf{u}}_i) \cdot \mathbf{K}/\omega] \qquad (D.53)$$

becomes a left-handed "quark-like" boost. The tachyon dynamical equation is[165]

$$[\gamma^D \partial/\partial t + \gamma \cdot (\nabla_r + i\nabla_i) - m]\psi_d(y) = 0 \qquad (D.54)$$

with the constraint equation

$$\nabla_r \cdot \nabla_i \, \psi_d(t, \mathbf{y}_r, \mathbf{y}_i) = 0 \qquad (D.55)$$

We will call the universe particles satisfying eqs. D.54 and D.55 left-handed *tachyon quark-like universe particles*.

D.3.1.8 Type IVb Case: Right-Handed Down-Quark-like Tachyon Universe Particles
   In this case we set $\hat{\mathbf{u}}_r \cdot \hat{\mathbf{u}}_i = 0$. Then by eq. D.7

---

[164] Here again the choice of ω in eq. 38.53 leads to a "left-handed" universe particle while the choice $\omega' = \omega - i\pi/2$ leads to a right-handed one.

[165] The gradient operators ∇_r and ∇_i are $(D - 1)$-dimensional spatial gradient operators.

$$\omega = (\omega_r^2 - \omega_i^2)^{\frac{1}{2}}$$

Thus ω again starts out either pure real (if $\omega_r \geq \omega_i$) or pure imaginary (if $\omega_r < \omega_i$). In this case we also choose ω real, and then change ω to

$$\omega = (\omega_r^2 - \omega_i^2)^{\frac{1}{2}} \rightarrow \omega' = (\omega_r^2 - \omega_i^2)^{\frac{1}{2}} - i\pi/2 = \omega - i\pi/2$$

since ω is a free parameter and proceed as we did in the prior case. The resulting Lorentz boost

$$\Lambda_C = \exp[i((\omega_r^2 - \omega_i^2)^{\frac{1}{2}} - i\pi/2)(\omega_r \hat{\mathbf{u}}_r + i\omega_i \hat{\mathbf{u}}_i) \cdot \mathbf{K}/\omega] \tag{D.56}$$

becomes a right-handed quark-like boost. The resulting tachyon dynamical equation is

$$[-\gamma^{\mathbf{D}} \partial/\partial t - \gamma \cdot (\nabla_r + i\nabla_i) - m]\psi_d(y) = 0 \tag{D.57}$$

with the constraint equation

$$\nabla_r \cdot \nabla_i \, \psi_d(t, \mathbf{y}_r, \mathbf{y}_i) = 0 \tag{D.58}$$

We will call the universe particles satisfying eqs. D.57 and D.58 right-handed *tachyon quark-like universe particles.*

D.3.2 Lagrangians

In this section we will develop a lagrangian formalism for each of the four types of fermion universe particles noting that a tachyon universe particles have two forms: left-handed and right-handed (discussed later in section D.3.5).

The various types of universe particles described in section D.3.1 correspond to universes with differing internal characteristics and motion in the Megaverse. The equations are all free field equations. Internal potentials and interactions must be introduced in these equations to complete the universe dynamical equations. A connection to the Wheeler-DeWitt description of their internal quantum structure also remains to be established (section D.3.6).

In defining the lagrangians for the four fermion universe types that yield their dynamical equations in a canonical manner, we require the conventional quantum field theory feature that the hamiltonian derived from the lagrangian is hermitean. We will develop a separate lagrangian for each type.

D.3.2.1 Type I Universe Particle Lagrangian

The Universe particle Dirac equation lagrangian is

$$\mathcal{L}_u = \bar{\psi}(i\gamma^\mu \partial/\partial y^\mu - m)\psi(y) \tag{D.59}$$

where

$$\bar{\psi} = \psi^\dagger \gamma^D$$

and $\psi^\dagger$ is the hermitean conjugate of ψ.

D.3.2.2 Type II Tachyon Universe Particle Lagrangian

This lagrangian includes both left-handed and right-handed cases. It can be separated into lagrangian terms for each case using parity projection operators.

$$\mathcal{L}_{uT} = \psi_T^{\;S}(\gamma^\mu \partial/\partial y^\mu - m)\psi_T(y) \tag{D.60}$$

where

$$\psi_T^{\;S} = \psi_T^{\;\dagger} \, i\gamma^D\gamma^5 \tag{D.61}$$

with γ^5 being the D-dimensional equivalent for γ^5 in 4 dimensions. The peculiar form of the tachyon universe lagrangian is necessitated by the hermiticity of the hamiltonian calculated from it.

D.3.2.3 Type III "Up-Quark-like" Universe Particle Lagrangian

The lagrangian density of a free "up-quark-like" universe particle is

$$\mathcal{L}_u = \bar{\psi}_u(i\gamma^\mu D_\mu - m)\psi_u(y) \tag{D.62}$$

where $\bar{\psi}_u = \psi_u^\dagger \gamma^D$ and

$$\psi_u^\dagger = [\psi_u(\mathbf{y_r}, \mathbf{y_i})]^\dagger \big|_{\mathbf{y_i} = -\mathbf{y_i}} \tag{D.63}$$

$$D_D = \partial/\partial y^D$$

$$D_k = \partial/\partial y_r{}^k + i\, \partial/\partial y_i{}^k \tag{D.64}$$

for $k = 1, 2, \ldots, (D-1)$. The action

$$I = \int d^{(D-1)}y\, \mathcal{L}_u \tag{D.65}$$

It is easy to show that this action is also real.

D.3.2.4 Type IV "Down-Quark-like" Tachyon Universe Particle Lagrangian

The lagrangian density of a free "down-quark-like" universe particle is

$$\mathcal{L}_d = \psi_d{}^C(y)(\gamma^\mathbf{D}\partial/\partial t + \gamma\cdot(\nabla_r + i\nabla_i) - m)\psi_d(y) \tag{D.66}$$

where

$$\psi_d{}^C(y) = [\psi_d(y)]^\dagger \big|_{\mathbf{y_i} = -\mathbf{y_i}} i\gamma^\mathbf{D}\gamma^5 \tag{D.67}$$

In words, eq. D.67 states: take the hermitean conjugate of $\psi_d(y)$; change $\mathbf{y_i}$ to $-\mathbf{y_i}$; and then post-multiply by the indicated factors.

The action is

$$I = \int d^{(D-1)}y\, \mathcal{L}_d \tag{D.68}$$

The action is real. The lagrangian can also be separated into left-handed and right-handed parts using projection operators.

D.3.3 Form of The Megaverse Quantum Coordinates Gauge Field

The discussions of sections D.3.1 and D.3.2 assumed the coordinates were Megaverse coordinates and their derivatives. Prior to those discussions we indicated we would use quantum coordinates in the Megaverse of the form[166]

$$Y^i(y) = y^i + i\ Y_u^{\ i}(y)/M_u^{D/2} \tag{D.4}$$

and their derivatives

$$\partial_i = \partial/\partial Y^i(y) = \partial/\partial(y^i - Y_u^{\ i}(y)/M_u^{D/2}) \tag{D.5}$$

for i = 1, 2, … , D to eliminate divergences in quantum field theory. The subscript "u" signifies universes. The mass constant for the Megaverse, M_u, may be the same as the mass constant M_c appearing in the Two-Tier mechanism for our universe.

In this section we define the gauge fields $Y_u^{\ i}(y)$ of the Megaverse.[167] They are similar to the $Y^\mu(y)$ fields of our New Standard Model.[168] The $Y_u(y)$ D-dimensional vector gauge field, in the absence of external sources, will be defined in a D-dimensional Coulomb gauge:

$$Y_u^{\ D}(y) = 0 \tag{D.69}$$
$$\partial Y_u^{\ j}(y)/\partial y^j = 0$$

where the sum over j is over the D − 1 spatial y coordinates. We follow a procedure similar to Blaha (2003) but for D-dimensional space. The lagrangian density for the free $Y_u^{\ j}(y)$ fields is

$$\mathscr{L}_u = -\tfrac{1}{4}\ F_u^{\ \mu\nu} F_{u\mu\nu} \tag{D.70}$$

and the lagrangian is

$$L_u = \int d^{(D-1)}y\ \mathscr{L}_u \tag{D.71a}$$

with

$$F_{u\mu\nu} = \partial Y_{u\mu}/\partial y^\nu - \partial Y_{u\nu}/\partial y^\mu \tag{D.71b}$$

---

[166] The denominator $M_u^{D/2}$ is necessitated by the dimension of $Y_u^{\ i}(y)$ which is $[m]^{D/2-1}$. Eqs. 38.78 and 38.81 below imply this conclusion.
[167] This choice implies that the Megaverse Y mass $M_u = M_C$, its universe mass.
[168] See Blaha (2005a) for details.

The equal time commutation relations, derived in the usual way, are:

$$[Y_u^\mu(\mathbf{y}, y^0), Y_u^\nu(\mathbf{y}', y^0)] = [\pi_u^\mu(\mathbf{y}, y^0), \pi_u^\nu(\mathbf{y}', y^0)] = 0 \qquad (D.72)$$

$$[\pi_u^j(\mathbf{y}, y^0), Y_{uk}(\mathbf{y}', y^0)] = -i\,\delta^{(D-1)tr}{}_{jk}(\mathbf{y} - \mathbf{y}') \qquad (D.73)$$

for $\mu, \nu, j, k = 1, 2, \ldots, (D-1)$ where

$$\pi_u^k = \partial\mathscr{L}_u/\partial Y_{uk}' \qquad (D.74)$$

$$\pi_u^0 = 0 \qquad (D.75)$$

and

$$\delta^{tr}{}_{jk}(\mathbf{y} - \mathbf{y}') = \int d^{(D-1)}k\; e^{i\,\mathbf{k}\bullet(\mathbf{y}-\mathbf{y}')}\,(\delta_{jk} - k_jk_k/\mathbf{k}^2)/(2\pi)^{D-1} \qquad (D.76)$$

$$Y_{uk}' = \partial Y_{uk}/\partial y^D \qquad (D.77)$$

The Coulomb gauge indicates $D-2$ degrees of freedom are present in the vector potential. The Fourier expansion of the vector potential is:

$$Y_u^i(y) = \int d^{(D-1)}k\; N_0(k) \sum_{\lambda=1}^{D-2} \varepsilon^i(k, \lambda)[a(k,\lambda)\,e^{-ik\cdot y} + a^\dagger(k,\lambda)\,e^{ik\cdot y}] \qquad (D.78)$$

where

$$N_0(k) = [(2\pi)^{(D-1)}2\omega_k]^{-\frac{1}{2}} \qquad (D.79)$$

and (since the field is massless)

$$k^D = \omega_k = (\mathbf{k}^2)^{\frac{1}{2}} \qquad (D.80)$$

where k^D is the energy, and where the $\varepsilon^i(k, \lambda)$ are the polarization unit vectors for $\lambda = 1,$ $\ldots, (D-2)$ and $k^\mu k_\mu = k^{D\,2} - \mathbf{k}^2 = 0$.

The commutation relations of the Fourier coefficient operators are:

$$[a(k,\lambda), a^\dagger(k',\lambda')] = \delta_{\lambda\lambda'}\delta^{(D-1)}(\mathbf{k} - \mathbf{k}') \qquad (D.81)$$

$$[a^\dagger(k,\lambda), a^\dagger(k',\lambda')] = [a(k,\lambda), a(k',\lambda')] = 0 \qquad (D.82)$$

and the polarization vectors satisfy

$$\sum_{\lambda=1}^{D-2} \varepsilon_i(k, \lambda)\varepsilon_j(k, \lambda) = (\delta_{ij} - k_ik_j/\mathbf{k}^2) \tag{D.83}$$

It will be convenient to divide the Y field into positive and negative frequency parts:

$$Y_u{}^+{}_i(y) = \int d^{(D-1)}k \, N_0(k) \sum_{\lambda=1}^{D-2} \varepsilon_i(k, \lambda) \, a(k,\lambda) \, e^{-ik\cdot y} \tag{D.84}$$

and

$$Y_u{}^-{}_i(y) = \int d^{(D-1)}k \, N_0(k) \sum_{\lambda=1}^{D-2} \varepsilon_i(k, \lambda) \, a^\dagger(k,\lambda) \, e^{ik\cdot y} \tag{D.85}$$

For later use we note the commutator between the positive and negative frequency parts is:

$$[\, Y_u{}^-{}_j(y_1), Y_u{}^+{}_k(y_2)] = -\int d^{(D-1)}k \, e^{ik\cdot(y_1 - y_2)} \, (\delta_{jk} - k_jk_k/\mathbf{k}^2)/[(2\pi)^{D-1} 2\omega_k] \tag{D.86}$$

D.3.3.1 Y^μ Fock Space Imaginary Coordinate States

States can also be defines for the quantized Y^μ field. These states will be similar in form to electromagnetic photon states but play a different role in our approach since they are in fact coordinate excitation states for the imaginary part of $Y^i(y)$ (eq. D.4). Thus universe particles (and other fields) will exist in a real D-dimensional space with quantum excitations into imaginary Quantum Dimensions. These excitations become significant at high energies. At low energies space appears as c-number complex; at very high energies space becomes slightly q-number complex.

There are two types of imaginary coordinate excitations: 1.) Quantum excitations into Fock states consisting of a superposition of states with a definite finite number of Y_u "particles" and 2.) Imaginary coordinate excitations into coherent Y_u states with an "infinite" number of particles. Coherent states can be viewed as representing "classical" fields.

In this section we will consider Y_u field states with a definite number of excitations ("particles"). The raising and lowering operators of the Y_u field can be used to define free particle states. For example a one particle state can be defined by

$$|k, \lambda> = a^\dagger(k, \lambda)|0> \qquad\qquad (D.87)$$

with corresponding bra state

$$<k, \lambda| = <0|a(k, \lambda) \qquad\qquad (D.88)$$

where the "coordinate vacuum" is defined as usual:

$$a(k, \lambda)|0> = 0 \qquad\qquad (D.89)$$
$$<0|a^\dagger(k, \lambda) = 0 \qquad\qquad (D.90)$$

Multi-particle states can also be defined in the conventional way with products of the raising and lowering operators applied to the vacuum. The set of all states containing a finite number of "particles" constitutes a Fock space.

A state with a finite number of Y_u "particles" represents a quantum fluctuation into imaginary Quantum Dimensions.

D.3.3.2 Y_u Coherent Imaginary Coordinate States

Coherent Y_u states bring us closer what we might consider to be "classical" imaginary dimensions – dimensions that we can, in principle, experience as we do normal dimensions. Let us define the coherent state[169]

$$| y, p> = e^{-\mathbf{p}\cdot \mathbf{Y_u}^-(y)/M_u{}^{D/2}}|0> \qquad\qquad (D.91)$$

This state is an eigenstate of the coordinate operator $Y_u{}^+(y')$:

---

[169] Coherent states are well known in the physics literature. See for example T. W. B. Kibble, J. Math. Phys. **9**, 315 (1968) and references therein; V. Chung, Phys. Rev. **140**, B1110 (1965); J. R. Klauder, J. McKenna, and E. J. Woods, J. Math. Phys. **7**, 822 (1966) and references therein.

$$Y_{u\ j}^{+}(y_1)\ |y_2, p> = -[Y_{u\ j}^{+}(y_1),\ \mathbf{p}\cdot\mathbf{Y}^{-}(y_2)]/M_u^{D/2}|y, p> \qquad (D.92)$$

$$= -\int d^{D-1}k\ [N_0(k)]^2\ e^{ik\cdot(y_2-y_1)}\ (p_j - k_j\mathbf{p}\cdot\mathbf{k}/\mathbf{k}^2)/M_u^{D/2}|y, p>$$

$$= p^i\Delta_{Tij}(y_1-y_2)/M_u^{D/2}|y, p> \qquad (D.93)$$

where $p^i\Delta_{Tij}(y_1 - y_2)/M_u^{D/2}$ is the eigenvalue of $Y_{u\ j}^{+}(y_1)$. As we will see later, the eigenvalue of Y_u^{+} becomes large as $(y_1 - y_2)^2 \to 0$. Thus the imaginary Quantum Dimensions become significant at very short distances, and then significantly modifies the high-energy behavior of quantum field theories. In particular, Quantum Dimensions have a significant effect when

$$(y_1 - y_2)^2 \lessgtr (2^{D-2}\pi^{D-2}M_u^2)^{-1} \qquad (D.94)$$

We assume the mass scale $M_u = M_C$ is very large – perhaps of the order of the Planck mass (1.221×10^{19} GeV/c^2).

D.3.3.3 Quantization of the Type I Free Universe Particle Dirac Field

The quantization procedure is formally identical to that of a conventional Dirac particle. The standard equal time anti-commutation relations for a D-dimensional fermion field are:

$$\{\psi_\alpha(Y),\ \psi_\beta(Y')\} = \{\pi_{\psi\alpha}(Y),\ \pi_{\psi\beta}(Y')\} = 0 \qquad (D.95)$$
$$\{\pi_{\psi\alpha}(Y),\ \psi_\beta(Y')\} = i\,\delta_{\alpha\beta}\,\delta^{D-1}(\mathbf{Y} - \mathbf{Y}') \qquad (D.96)$$

where α and β are the spinor indices ranging from 1 to $N_{MRC} = 2^{D/2}$ and where

$$\pi_{\psi\alpha}(Y) = i\,\psi_\alpha^{\dagger}(Y) \qquad (D.97)$$

The field can be expanded in a fourier series:

$$\psi(Y(y)) = \sum_s \int d^{D-1}p \; N^d_m(p) \; [b(p,s)u(p,s) :e^{-ip\cdot(y+iYu/M_u^{D/2})}: +$$

$$+ d^\dagger(p,s)v(p,s) :e^{ip\cdot(y+iYu/M_u^{D/2})}:] \tag{D.98}$$

$$\psi^\dagger(Y(y)) = \sum_s \int d^{D-1}p \; N^d_m(p) \; [b^\dagger(p,s)\bar{u}(p,s)\gamma^0 :e^{+ip\cdot(y+iYu/M_u^{D/2})}: +$$

$$+ d(p,s)\bar{v}(p,s)\gamma^0 :e^{-ip\cdot(y+iYu/M_u^{D/2})}:] \tag{D.99}$$

where

$$N^d_m(p) = [m/((2\pi)^{D-1}E_p)]^{1/2} \tag{D.100}$$

and

$$E_p = p^D = (\mathbf{p}^2 + m^2)^{1/2} \tag{D.101}$$

with : ... : signifying normal ordering. The commutation relations of the Fourier coefficient operators are:

$$\{b(p,s), b^\dagger(p',s')\} = \delta_{ss'}\delta^{D-1}(\mathbf{p}-\mathbf{p}') \tag{D.102}$$

$$\{d(p,s), d^\dagger(p',s')\} = \delta_{ss'}\delta^{D-1}(\mathbf{p}-\mathbf{p}') \tag{D.103}$$

$$\{b(p,s), b(p',s')\} = \{d(p,s), d(p',s')\} = 0 \tag{D.104}$$

$$\{b^\dagger(p,s), b^\dagger(p',s')\} = \{d^\dagger(p,s), d^\dagger(p',s')\} = 0 \tag{D.105}$$

$$\{b(p,s), d^\dagger(p',s')\} = \{d(p,s), b^\dagger(p',s')\} = 0 \tag{D.106}$$

$$\{b^\dagger(p,s), d^\dagger(p',s')\} = \{d(p,s), b(p',s')\} = 0 \tag{D.107}$$

The spinors $u(p,s)$ and $v(p,s)$ are defined in a conventional way (as in Bjorken and Drell). However their form is different from the 4-dimensional case. If one takes the $N_{MRC}\times N_{MRC} \equiv 2^{D/2}\times 2^{D/2}$ $\gamma\cdot p$ matrix, then the first $2^{D/2-1}$ columns give $u(p,s)$ up to a normalization for the free particle case, the remaining $2^{D/2-1}$ columns give $v(p,s)$ up to a normalization.

Since there are $2^{D/2-1}$ possible spin values, using the equation $2s+1 =$ total number of spin values, we see that the spin of a fermion universe particle is

$$s_M = 2^{D/2-2} - \tfrac{1}{2}$$

The possible universe particle spin values are:

Up spin values: $+1/2^{D/2-1}$, $+2/2^{D/2-1}$, ... , $+2^{D/2-2}/2^{D/2-1}$
Down spin values: $-2^{D/2-2}/2^{D/2-1} = -\frac{1}{2}$, $-2^{D/2-2}/2^{D/2-1} + 1$, ... , $-1/2^{D/2-1}$

This enormous number of possible spins is reasonable considering the number of dimensions D and the enormous variety in spins one should expect in universes – given their large size and complexity.

D.3.3.4 Feynman Propagators for the Type I Free Universe Particle Dirac Field

The form of the fermion universe particle Feynman propagator differs from a conventional fermion propagator by having a Gaussian factor $R(\mathbf{p}, z)$ in its fourier expansions. This follows from using quantum Megaverse coordinates (eq. D.4).

$$iS_F^{TT}(y_1 - y_2) = \langle 0|T(\overline{\psi}(Y(y_1))\psi(Y(y_2)))|0\rangle \qquad (D.108)$$

where the time ordering is with respect to y_1^D and y_2^D. Expanding the free fields leads to the fourier representation:

$$iS_F^{TT}(y_1 - y_2) = i \int \frac{d^D p\, e^{-ip\cdot(y_1-y_2)}\, (\mathbf{p}+ m)\, R(\mathbf{p}, y_1 - y_2)}{(2\pi)^D\, (p^2 - m^2 + i\varepsilon)} \qquad (D.109)$$

Where

$$R(\mathbf{p}, y_1 - y_2) = \exp[-p^i p^j \Delta_{Tij}(y_1 - y_2)/M_u^D] \qquad (D.110)$$
$$= \exp\{-p^2[A(v) + B(v)\cos^2\theta] / [(2\pi)^{D-2} M_c^4 z^2]\} \qquad (D.111)$$

(Note p^2 is the square of the spatial $(D-1)$-vector.) with

$$z^\mu = y_1^\mu - y_2^\mu \qquad (D.112)$$
$$z = |\mathbf{z}| = |\mathbf{y_1} - \mathbf{y_2}| \qquad (D.113)$$

$$p = |\mathbf{p}| \tag{D.114}$$
$$v = |z^0|/z \tag{D.115}$$
$$A(v) = (1 - v^2)^{-1} + .5v \ln[(v - 1)/(v + 1)] \tag{D.116}$$
$$B(v) = v^2(1 - v^2)^{-1} - 1.5v \ln[(v - 1)/(v + 1)] \tag{D.117}$$
$$\mathbf{p}\cdot\mathbf{z} = pz \cos\theta \tag{D.118}$$

and $|\mathbf{p}|$ denoting the length of a spatial $(D - 1)$-vector $\mathbf{p}$ while $|z^0|$ is the absolute value of $z^0 \equiv z^D$.

As eq. D.109 indicates, the Gaussian damping factor[170] $R(p, z)$ for large spatial momentum p is the same for both the positive and negative frequency parts of the Two-Tier Feynman propagator. We are assuming the spatial momentum is real-valued in this discussion. It is also important to note that $R(p, z)$ does not depend on $p^0 = p^D$ (in the Y Coulomb gauge) and thus the integration over p^0 proceeds in the usual way to produce time-ordered positive and negative frequency parts.

D.3.3.5 Feynman Propagators for the Types II, III, and IV Free Universe Particle Dirac Fields

These propagators differ in details from the Type I propagator. The differences modulo the change in dimension appear in Blaha (2011c). See also Blaha (2005a) for a detailed discussion of 4-dimensional spin ½ particle propagators.

D.3.4 Expanding and Contracting Universes: Impact of Time Dependent Universe Particle Masses

Our discussions of the dynamics of universe particles assumed their masses were constant. However the definition of mass in terms of the area of a universe based on the physics of black holes is

$$M = \kappa A/8\pi \tag{D.119}$$

---

[170] Note the Gaussian damping is for all $D - 1$ spatial momentum integrations.

where A is the area of the black hole shows that *the mass of a universe particle is time dependent* because the area of a universe is generally time dependent. For example, our universe is expanding and its surface area is thus growing with time.

Eqs. D.11 (and subsequent fermion dynamic equations) must then be modified from

$$\psi(y) = \exp[-imt]w(0) \tag{D.11}$$

to a covariant form:

$$\psi(y) = \exp[-i\int_0^{w\cdot y} m(t')dt']w(0) \tag{D.120}$$

where w is a unit D-vector in the time (y^D) direction ($w^2 = 1$). The lower bound on the integral, 0, is the time of the beginning of the universe particle – its Big Bang. Thus the cumulative change in the mass of the universe particle may be significant. It is interesting to note that the Wheeler-Dewitt equation also has a variable value mass term R that also depends on the evolution of the universe.

Eq. D.120 satisfies the free covariant Dirac-like universe particle field dynamic equation

$$[i\gamma^i\partial/\partial y^i - m(w\cdot y)]\psi(y) = 0 \tag{D.121}$$

In contrast to the constant mass equation eq. D.19. Substituting eq. D.120 in eq. D.121 we find

$$(\gamma^i w_i \, m(w\cdot y) - m(w\cdot y))\psi(y) = 0 \tag{D.122}$$

or

$$(\gamma^i w_i - 1)\psi(y) = 0 \tag{D.123}$$

Upon performing a D-dimensional Lorentz boost (of the type of eqs. D.13 – D.16) on eq. D.123 we obtain

$$(\gamma_i p^i/m_0 - 1)\psi(y) = 0$$

or

$$(\gamma_i p^i - m_0)\psi(y) = 0 \tag{D.124}$$

where p^i is a momentum D-vector with $p^2 = m_0^2$. Eq. D.123 is the constant mass momentum space dynamic equation. It determines the spinor in $\psi(y)$. After taking account of the quantum coordinates the quantum Dirac-like universe particle wave function has the form

$$\psi(Y(y)) = \sum_s \int d^{(D-1)}p \, N^d_m(p) \, [b(p,s)u(p,s) : \exp[-iG(p, Y(y))]: + d^\dagger(p,s)v(p,s) \cdot$$
$$\cdot :\exp[+iG(p, Y(y))]:\} \tag{D.125}$$

$$\psi^\dagger(Y(y)) = \sum_s \int d^{(D-1)}p \, N^d_m(p) \, \{b^\dagger(p,s)\bar{u}(p,s)\gamma^0 :\exp[+iG(p, Y(y))]: + d(p,s)\bar{v}(p,s)\gamma^0 \cdot$$
$$\cdot \exp[-iG(p, Y(y))]:\} \tag{D.126}$$

where : ... : denotes normal ordering and

$$G(p, Y(y)) = \int_0^{p\cdot Y(y)/\lambda} m(t')dt' \tag{D.127}$$

with $\lambda = m_0$, and $N^d_m(p)$ a normalization constant. Contrast eqs. D.125-D.126 to the constant mass case eqs. D.98-D.101. The *constant mass case* simply sets $m(t') = m_0$.

If we examine the integral eq. D.127 for a short time interval δt in the particle's rest frame then $G(p, Y(y)) \approx m(0)\delta t$ and so we define $m(0) = m_0$. Based on the formula for universe particle mass (eq. D.119) we anticipate that m_0 might be as large as the Planck mass or larger – thus an extremely short radius. Blaha (2013) describes a quantum Big Bang model in which the initial radius of the universe is $O(EM_{Planck}^{-2})$ where E is of the order of 1 and has the dimensions of [mass].

Thus we have a closed form definition of a quantum universe particle wave function for universe particles of type I. A similar procedure can be followed for universe particles of types II, III, and IV.

The Feynman propagator for type I quantum fields is *not* eq. D.109 but now has a form reflecting the $Y(y)$ dependence of the quantum fields in eqs. D.125 and D.126:

$$iS_F^{TT}(y_1, y_2) = i \int \frac{d^D p \; \{ <0|\theta(y_{1D} - y_{2D})G(y_1, y_2) + \theta(y_{2D} - y_{1D})G(y_2, y_1)\}0>}{(2\pi)^D (p - m_0)} \qquad (D.128)$$

where p^D is the energy and

$$G(y_1, y_2) = : \exp[-iG(p, Y(y_1))]: :\exp[+iG(p, Y(y_2))]: \qquad (D.129)$$

Let

$$G_{tot}(y_1, y_2) = <0|\theta(y_{1D} - y_{2D})G(y_1, y_2) + \theta(y_{2D} - y_{1D})G(y_2, y_1)\}0> \qquad (D.130)$$
$$= <0|\theta(y_{1D} - y_{2D}):\exp[-iG(p, Y(y_1))]::\exp[+iG(p, Y(y_2))]: +$$
$$+ \theta(y_{2D} - y_{1D}) :\exp[-iG(p, Y(y_2))]::\exp[+iG(p, Y(y_1))):]|0>$$

$$= <0|\theta(y^D_1 - y^D_2): \exp[-i\int_0^{p \cdot Y(y1)/\lambda} m(t')dt']::\exp[+i\int_0^{p \cdot Y(y2)/\lambda} m(t')dt']: +$$

$$+ \theta(y^D_2 - y^D_1):\exp[-i \exp[+i\int_0^{p \cdot Y(y2)/\lambda} m(t')dt']::\exp[+i\int_0^{p \cdot Y(y1)/\lambda} m(t')dt']:|0>$$

with $\lambda = m_0$ then

$$iS_F^{TT}(y_1, y_2) = i \int \frac{d^D p \; G_{tot}(y_1, y_2)}{(2\pi)^D (p - m_0)} \qquad (D.131)$$

Except for the case of a constant mass, where $m(t) = m_0$, the Feynman propagator is not a function of $y_1 - y_2$. The evaluation of eq. D.130 in the general case of a variable mass is straightforward but cumbersome. For the special case of a linear time dependence of the mass, $m(t) = at$, we find eq. D.130 gives

$$G_{tot}(y_1, y_2) = <0|\theta(y^D{}_1 - y^D{}_2):\exp[-ia(p \cdot Y(y_1)/m_0)^2/2]::\exp[+ia(p \cdot Y(y_2)/m_0)^2/2]: + $$
$$+ \theta(y^D{}_1 - y^D{}_2):\exp[-ia(p \cdot Y(y_2)/m_0)^2/2]::\exp[+ia(p \cdot Y(y_1)/m_0)^2/2]:|0>$$

(D.132)

yielding a complex function of p, y_1, and y_2. *Note that the lower bound of the integrals in the Feynman propagator cancel and thus the need for an understanding of the beginning of a universe is removed in this case.*

We have shown that universe particle theory can handle the case of a variable universe mass $m(t)$. Expanding or contracting (or oscillating) universe particles correspond to expanding and contracting (or oscillating) universes.

D.3.5 Left-Handed and Right-handed Universe Particles

In sections D.3.1 and D.3.2 we found that left-handed and right-handed tachyon universe particles existed. The tachyon nature of the universe particles indicates that their speed in the universe exceeds the "speed of light" of the Megaverse. The physical meaning of the handedness of these types of universes is an interesting issue. When we consider our universe we see left-handedness in the weak interactions of elementary particles. In addition it appears that organic molecules overwhelmingly favor left-handedness on earth although right-handed molecules exist in outer space and can be created in the laboratory. Right-handed molecules transform into left-handed molecules in watery media through electromagnetic effects.

Why nature favors left-handedness is an open question. It has given rise to speculations that gravitation, especially quantum gravitons, may be left-handed. The European Space Agency's Planck telescope will study polarization effects in the cosmos and may well be able to show that the gravitons starting from the beginning of the universe, and magnified by inflation in the universe's expansion, may be left-handed.

If handedness of gravitation is verified experimentally, then our theory of left-handed/right-handed universe particles would be supported. *Our universe would then be tachyon and probably left-handed. We, in the universe, would, of course, not know of the velocity of the universe in the Megaverse.*

D.3.6 Internal Structure of Universe Particles

We have treated universes as particles in the preceding discussion taking an extremely large view of Megaverse particles just as elementary particle theory viewed nucleons at low energies (large distances). Now we develop a more detailed view of universe particles in a manner analogous to the high energy view of the internal dynamics of nucleons that led to the quark-parton model of nucleons. In the present case we shall see that high energy Baryonic and other field probes of universe particles can yield a model of the internal structure of universes.

We know that universes are composed of matter and radiation. We believe that there is at least one possible accessible interaction between universes dependent on baryon number – a baryonic, D-dimensional gauge field. There are also other particle number gauge fields – but these are less likely to be significant since Dark matter has yet to be found except through its gravitational effects. In this section we will discuss the use of a baryonic gauge field to probe the baryon structure of universe particles.

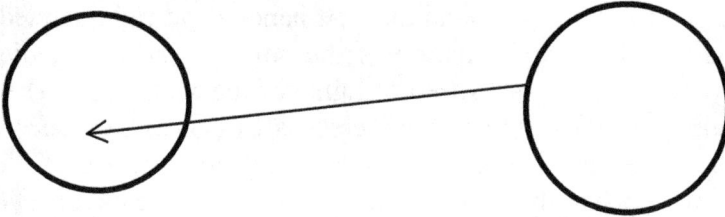

Figure D.1. A symbolic view of a high resolution (high energy) probe from a universe to a specific baryonic part of another universe.

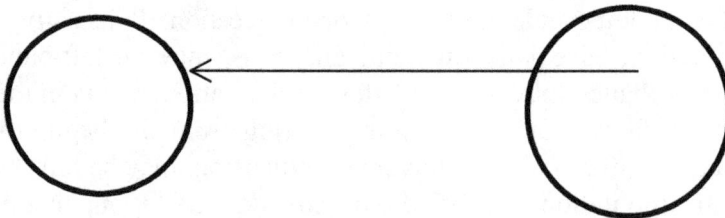

Figure D.2. A symbolic view of a low resolution (lower energy) probe from a universe to an entire universe.

There appears to be two types of probes[171] of a universe: 1) a series of high energy probes to specific small regions inside another universe for the purpose of mapping its internal structure; 2) a low energy probe of another universe to get a global view of its structure.[172] The first type of probe corresponds to deep inelastic (high energy) electron-nucleon scattering which led to the quark-parton model of nucleons. The second type of probe corresponds to low energy electron-nucleon scattering to get a "global" view of a nucleon. In both case an electromagnetic (gauge) field particle (photon) was the probe particle.

Besides the inherent scientific interest in such experiments it is possible that they may be of use in the very distant future if Mankind is able to develop Megaverse starships that can travel in the Megaverse to other universes. Then the baryonic gauge field may become the "eyes" of the starship in addition to the electromagnetic field (light) are the eyes of current spaceships. We considered the possibility of universe starships in the book entitled, *All the Megaverse! II* in detail.

D.4 Boson Universe Particles

The previous sections has described fermion universe particles. In this section we will briefly describe aspects of boson (spin 0) universe particles. First it is important to note that the Wheeler-DeWitt equation being second order like the Klein-Gordon equation seems to suggest that universe particles can be bosonic – like Klein-Gordon equation particles.[173] The Wheeler-DeWitt equation has a mass-like term R that can be positive or negative. If the mass term is negative then the wave-like propagation of the state functional (wave equation solution) can be in space-like directions implying a tachyon solution. Thus the Wheeler-DeWitt equation supports "normal" state functionals that propagate in time-like directions as well as tachyon propagation.

---

[171] It would appear that the probes are in Megaverse coordinates, not universe coordinates, in order to 'bridge' the distance between universes. Chapter 35 shows the manner of the mapping between Megaverse coordinates and universe coordinates for quantum fields.

[172] **We note these probes, being limited by the speed of light, are at best theoretical speculations since the travel time between universes is so very large.**

[173] One should remember that the Wheeler-DeWitt equation is not in space but in a 6-dimensional manifold, denoted M, of metrics with one "time" dimension – having hyperbolic signature $- + + + + +$ when the metric is positive definite. See DeWitt's paper.

For this reason we suggest that boson universe particles can be either normal or tachyonic. Tachyon bosonic universe particles can fission in a manner similar to tachyonic fermion universe particles. The fission equations of section D.5.2 also apply to tachyonic boson universe particles.

The quantum field theory of normal and tachyonic boson universe particles is similar to that of ordinary bosons. See Blaha (2005a) for the boson case discussion that is paralleled by our universe particle formalism.

D.5 Physical Meaning of Universe Particle Spin

The physical meaning of spin is a continuing discussion topic. We have suggested[174] that spin states are in essence logic states with changes in spin an analogous to changes in logical values in a discourse or computer program. Since the matrix formalism for spin ½ and higher spin states is formally similar to the formalism for angular momentum, one can combine spin and angular momentum as we do in quantum theory.

In the case of universe particles, one can also associate universe particles with "true" and "false" values. Fermion universe states have $2^{D/2 - 1}$ 'truth' values and correspond to a multi-valued logic. The numerousness of 'truth' values is due to the D-dimensional space within which universe particles reside.

Naturally one would like to know the physical differences between these $2^{D/2 - 1}$ types of universe particles. Does the difference reside in different shapes of the universe particles? Or is the difference somehow a consequence of the global mass-energy distribution of the universe that we have not been able to discern since we only know of one universe?

The physical meaning of spin for elementary particles is also somewhat elusive. It does not reflect the flow of charge within a particle. For if it did reflect physical spinning of a particle, the outer edges of a particle such as an electron would be traveling at a speed faster than light. So spin is not a mechanical property of the internal structure of an elementary particle. We have suggested that it is a truth value (within a

---

[174] Blaha (2011c) and subsequent books.

particle core) in the matrix formulation of a 4-valued logic called Asynchronous Logic. Thus it has no certain tangible physical basis.

In the case of universe particles the situation is unclear at present. It could be taken to be an indirect reflection of the structure of mass-energy within a universe. This view would be contrary to our proposed view of elementary particle spin as truth values. So we can only assert that a logic interpretation is the only sensible one (based on our present knowledge or our lack thereof). The physical role of universe particle spin is only evident in interactions between universe particles via gauge fields. Thus one must simply view it as a construct for the present.

Appendix E. Megaverse Interactions

E.1 Unified SuperStandard Theory Interactions in the Megaverse

The particles and interactions of the Unified SuperStandard Theory exist in the Megaverse space between universes. They are, of course, required to be specified in terms of Megaverse coordinates.

Boson representations are described in Megaverse coordinates in sections 33.3.2, 34.3.6, chapter 35, and section 36.6.

Chapter 38, which describes fermion universe particles, also provides a description of fermion particle fields in the Megaverse.

E.2 Megaverse Lorentz Group

The Megaverse Lorentz group is SU(191,1) – a direct generalization of the SU(3, 1) Lorentz group for Special Relativity in our universe. The apparently extremely low density of the Megaverse suggests the Megaverse is most likely a flat space-time. In our work we have chosen y^D transformed to a real value as Megaverse time. This assumption is required in order for the Megaverse to be a dynamic entity. We assume the time variable is real since clocks measure real-valued time. A Megaverse Reality group transformation can map complex time coordinates to real values with no loss of generality.

E.3 Megaverse Gravitation

Megaverse Complex General Relativity and Gravitation is also a direct generalization of the 4-dimensional case with one significant exception: the Einstein field equations change to:

$$R_{\mu\nu} - [1/(D - 2)]g_{\mu\nu}R = -8\pi GT_{\mu\nu}$$

where the metric is such that $g_\mu{}^\mu = 190$, $R_{\mu\nu}$ is the Megaverse Ricci tensor, and R is the Megaverse curvature scalar. (The overall form of these quantities is the same as that of 4-dimensional space-time.)

E.4 Megaverse SuperSymmetry

The discussions of the Θ-symmetry and SuperSymmetry in our 4-dimensional space-time were presented earlier. A similar discussion of Θ-symmetry and SuperSymmetry with additional ramifications will now be considered. It was discussed to some extent in Blaha (2017c).

An important new feature relating to Θ-symmetry and SuperSymmetry is the dimension of the Megaverse D = 192. Due to the value of this dimension we can define a U(192) unitary Reality group that transforms the complex-valued coordinates of the Megaverse to real-valued coordinates. We also have the Θ-symmetry for interactions. Lastly we can define 192 antisymmetric coordinates that supports Supersymmetric field theory in the Megaverse.

This section discusses these possibilities. Much of the discussion parallels the discussions in Blaha (2017c).

E.4.1 Θ-SymmetryTransformations in the Megaverse

The choice of dimension, 192, enables us to extend the Θ-Symmetry to Megaverse coordinates. Let $C_\Theta(y)$ be a local U(4) Θ-Symmetry transformation that rotates interactions in Megarverse space y. Then applying a Θ-Symmetry transformation to the Dirac equation in Megaverse space:[175]

$$\gamma_\mu D^\mu \psi = \gamma_\mu \{\partial^\mu + i\, [g_\Theta A_\Theta{}^{1\mu}(y) + g_\Theta A_\Theta{}^{2\mu}(y) + \mathbf{A}_I{}^{1\mu}(y) + \mathbf{A}_I{}^{2\mu}(y)]\}\psi(y) = 0$$

we find it transforms to

$$\gamma_\mu \{\partial^\mu + i\, [g_\Theta A'_\Theta{}^{1\mu}(y) + g_\Theta A'_\Theta{}^{2\mu}(y) + \mathbf{A}'_I{}^{1\mu}(y) + \mathbf{A}'_I{}^{2\mu}(y)]\}\psi'(y') = 0$$

---

[175] We omit the Species group interaction.

where

$$A'_\Theta{}^{1\mu}(y') = C_\Theta(y)A_\Theta{}^{1\mu}(y)C_\Theta{}^{-1}(y) - i\, C_\Theta(y)\partial^\mu C_\Theta{}^{-1}(y)/g_\Theta$$
$$A'_\Theta{}^{2\mu}(y') = C_\Theta(y)A_\Theta{}^{2\mu}(y)C_\Theta{}^{-1}(y)$$
$$\mathbf{A'}_I{}^{1\mu}(y') = C_\Theta(y)\mathbf{A}_I{}^{1\mu}(y)\, C_\Theta{}^{-1}(y)$$
$$\mathbf{A'}_I{}^{2\mu}(y') = C_\Theta(y)\mathbf{A}_I{}^{2\mu}(y)\, C_\Theta{}^{-1}(y)$$
$$\psi'(y') = C_\Theta(y)\psi(y)$$

The Θ-Symmetry interaction is experienced by all particles (fields) in the Megaverse. It is also experienced by all universes in the Megaverse and thus affects universe dynamics.

E.5 Megaverse SuperSpace

The Megaverse offers an exciting new platform for SuperSymmetry studies of the Unified SuperStandard Theory and possible SuperSymmetric generalizations using Megaverse Superspace with antisymmetric coordinates in addition to normal coordinates.

We can define a generalized set of Megaverse coordinates that associate 192 antisymmetric coordinates with the 192 normal Megaverse coordinates:

$$\{y^\mu, \Theta^\nu\}$$

where $\mu = 1, \dots , 192$ *and* also $\nu = 1, \dots , 192$. We now have an equal numbers of normal and antisymmetric coordinates in Megaverse superspace. This was not the case in 4-dimensional space-time.

Supersymmetric transformations on superspace coordinates have the infinitesimal form:

$$y^\mu \rightarrow y^\mu + i\, \varepsilon\, \Gamma^\mu \Theta$$
$$\Theta^\nu \rightarrow \Theta^\nu + \varepsilon^\nu$$

where Γ^μ is a 192×192 matrix generalization of the Dirac γ matrices.[176] There are 192 Γ matrices.

　　Using Megaverse superspace coordinates one can proceed to construct a superfields formalism with many features analogous to the 4-dimensional formulation but of significantly greater complexity. We defer doing this in view of the finiteness of Megaverse Two-Tier Quantum Field Theory which, like the 4-dimensional case, yields totally finite results in all orders of perturbation theory. We note the major benefit of 4-dimensional SuperField Theory was the finiteness of calculations in perturbation theory. Given the complexity of Megaverse SuperField perturbation theory it would appear the simplicity of Two-Tier Quantum Field Theory is preferable.

---

[176] See Blaha (2017c) for their definition for arbitrary dimension D.

Appendix F. "Approximate Calculation of the Eigenvalue Function in Massless Quantum Electrodynamics"

This refereed paper is S. Blaha, Phys. Rev. **D9**, 2246 (1974). Reprinted with the kind permission of Physical Review D.

REFERENCES

Akhiezer, N. I., Frink, A. H. (tr), 1962, *The Calculus of Variations* (Blaisdell Publishing, New York, 1962).

Bjorken, J. D., Drell, S. D., 1964, *Relativistic Quantum Mechanics* (McGraw-Hill, New York, 1965).

Bjorken, J. D., Drell, S. D., 1965, *Relativistic Quantum Fields* (McGraw-Hill, New York, 1965).

Blaha, S., 1998, *Cosmos and Consciousness* (Pingree-Hill Publishing, Auburn, NH, 1998).

_____, 2004, *Quantum Big Bang Cosmology: Complex Space-time General Relativity, Quantum Coordinates, Dodecahedral Universe, Inflation, and New Spin 0, ½, 1 & 2 Tachyons & Imagyons* (Pingree Hill Publishing, Auburn, NH, 2004).

_____, 2018e, *Unification of God Theory and Unified SuperStandard Model THIRD EDITION* (Pingree Hill Publishing, Auburn, NH, 2018).

_____, 2019a, *Calculation of: QED α = 1/137, and Other Coupling Constants of theUnified SuperStandard Theory* (Pingree Hill Publishing, Auburn, NH, 2019).

_____, 2019b, *Coupling Constants of the Unified SuperStandard Theory SECOND EDITION: We Find the Fine Structure Constant 1/137.0359801, and so: OUR UNIVERSE AND LIFE! Also a Universal Eigenvalue Function for all Known Interactions, And Running Coupling Constants to all Perturbative Orders* (Pingree Hill Publishing, Auburn, NH, 2019).

_____, 2019c, *New Hybrid Quantum Big_Bang-Megaverse_Driven Universe with a Finite Big Bang and an Increasing Hubble Constant* (Pingree Hill Publishing, Auburn, NH, 2019).

_____, 2019d, *The Universe, the Electron, and the Vacuum* (Pingree Hill Publishing, Auburn, NH, 2019).

Gelfand, I. M., Fomin, S. V., Silverman, R. A. (tr), 2000, *Calculus of Variations* (Dover Publications, Mineola, NY, 2000).

Gradshteyn, I. S. and Ryzhik, I. M., 1965, *Table of Integrals, Series, and Products* (Academic Press, New York, 1965).

Heitler, W., 1954, *The Quantum Theory of Radiation* (Claendon Press, Oxford, UK, 1954).

Huang, Kerson, 1992, *Quarks, Leptons & Gauge Fields 2nd Edition* (World Scientific Publishing Company, Singapore, 1992).

Kaku, Michio, 1993, *Quantum Field Theory*, (Oxford University Press, New York, 1993).

Landau, L. D. and Lifshitz, E. M., 1987, *Fluid Mechanics 2nd Edition*, (Pergamon Press, Elmsford, NY, 1987).

Misner, C. W., Thorne, K. S., and Wheeler, J. A., 1973, *Gravitation* (W. H. Freeman, New York, 1973).

Streater, R. F. and Wightman, A. S., 2000, *PCT, Spin, Statistics, and All That* (Princeton University Press, Princeton, NJ 2000).

Weinberg, S., 1972, *Gravitation and Cosmology* (John Wiley and Sons, New York, 1972).

Weinberg, S., 1995, *The Quantum Theory of Fields Volume I* (Cambridge University Press, New York, 1995).

Weinberg, S., 2000, *The Quantum Theory of Fields Volume III Supersymmetry* (Cambridge University Press, New York, 2000).

Weyl, H., 1950, *Space, Time, Matter* (Dover, New York, 1950).

INDEX

About the Author

Stephen Blaha is a well-known Physicist and Man of Letters with interests in Science, Society and civilization, the Arts, and Technology. He had an Alfred P. Sloan Foundation scholarship in college. He received his Ph.D. in Physics from Rockefeller University. He has served on the faculties of several major universities. He was also a Member of the Technical Staff at Bell Laboratories, a manager at the Boston Globe Newspaper, a Director at Wang Laboratories, and President of Blaha Software Inc. and of Janus Associates Inc. (NH).

Among other achievements he was a co-discoverer of the "r potential" for heavy quark binding developing the first (and still the only demonstrable) non-abelian gauge theory with an "r" potential; first suggested the existence of topological structures in superfluid He-3; first proposed Yang-Mills theories would appear in condensed matter phenomena with non-scalar order parameters; first developed a grammar-based formalism for quantum computers and applied it to elementary particle theories; first developed a new form of quantum field theory without divergences (thus solving a major 60 year old problem that enabled a unified theory of the Standard Model and Quantum Gravity without divergences to be developed); first developed a formulation of complex General Relativity based on analytic continuation from real space-time; first developed a generalized non-homogeneous Robertson-Walker metric that enabled a quantum theory of the Big Bang to be developed without singularities at t = 0; first generalized Cauchy's theorem and Gauss' theorem to complex, curved multi-dimensional spaces; received Honorable Mention in the Gravity Research Foundation Essay Competition in 1978; first developed a physically acceptable theory of faster-than-light particles; first derived a composition of extrema method in the Calculus of Variations; first quantitatively suggested that inflationary periods in the history of the universe were not needed; first proved Gödel's Theorem implies Nature must be quantum; provided a new alternative to the Higgs Mechanism, and Higgs particles, to generate masses; first showed how to resolve logical paradoxes including Gödel's Undecidability Theorem by developing Operator Logic and Quantum Operator Logic; first developed a quantitative harmonic oscillator-like model of the life cycle, and interactions, of civilizations; first showed how equations describing superorganisms also apply to civilizations. A recent book shows his theory applies successfully to the past 14 years of history and to *new* archaeological data on Andean and Mayan civilizations as well as Early Anatolian and Egyptian civilizations.

He first developed an axiomatic derivation of the form of The Standard Model from geometry – space-time properties – The Unified SuperStandard Model. It unifies all the known forces of Nature. It also has a Dark Matter sector that includes a Dark ElectroWeak sector with Dark doublets and Dark gauge interactions. It uses quantum coordinates to remove infinities that crop up in most interacting quantum field theories and additionally to remove the infinities that appear in the Big Bang and generate inflationary growth of the universe. It shows gravity has a MOND-like form without sacrificing Newton's Laws. It relates the interactions of the MOND-like sector of gravity with the r-potential of Quark Confinement. The axioms of the theory lead to the question of their origin. We suggest in the preceding edition of this book it can be attributed to an entity with God-like properties. We explore these properties in "God Theory" and show they predict that the Cosmos exists forever although individual universes (or incarnations of our universe) "come and go." Several other important results emerge from God Theory such a functionally triune God. The Unified SuperStandard Theory has many other

important parts described in the Current Edition of *The Unified SuperStandard Theory* **and expanded in subsequent volumes.**

In 2019 Blaha calculated the correct value of the QED Fine Structure Constant based on his 1973 vacuum polarization paper. He also calculated approximate values for Standard Model U(1), SU(2), and SU(3) interactions, He also extended his 2004 model of the quantum Big Bang to the entire expansion of the universe using a new universal scale factor formulation. This formulation appears to be analogous to QED vacuum polarization..

Blaha has had a major impact on a succession of elementary particle theories: his Ph.D. thesis (1970), and papers, showed that quantum field theory calculations to all orders in ladder approximations could not give scaling deep inelastic electron-nucleon scattering. He later showed the eigenvalue equation for the fine structure constant α in Johnson-Baker-Willey QED had a zero at $\alpha = 1$ not 1/137 by solving the Schwinger-Dyson equations to all orders in an approximation that agreed with exact results to 4th order in α thus ending interest in this theory. In 1979 at Prof. Ken Johnson's (MIT) suggestion he calculated the proton-neutron mass difference in the MIT bag model and found the result had the wrong sign reducing interest in the bag model. These results all appear in Physical Review papers. In the 2000's he repeatedly pointed out the shortcomings of SuperString theory and showed that The Standard Model's form could be derived from space-time geometry by an extension of Lorentz transformations to faster than light transformations. This deeper space-time basis greatly increases the possibility that it is part of THE fundamental theory. Recently, Blaha showed that the Weak interactions differed significantly from the Strong, electromagnetic and gravitation interactions in important respects while these interactions had similar features, and suggested that ElectroWeak theory, which is essentially a glued union of the Weak interactions and Electromagnetism, possibly modulo unknown Higgs particle features, be replaced by a unified theory of the other interactions combined with a stand-alone Weak interaction theory. Blaha also showed that, if Charmonium calculations are taken seriously, the Strong interaction coupling constant is only a factor of five larger than the electromagnetic coupling constant, and thus Strong interaction perturbation theory would make sense and yield physically meaningful results.

In graduate school (1965-71) he wrote substantial papers in elementary particles and group theory: The Inelastic E- P Structure Functions in a Gluon Model. Phys. Lett. B40:501-502,1972; Deep-Inelastic E-P Structure Functions In A Ladder Model With Spin 1/2 Nucleons, Phys.Rev. D3:510-523,1971; Continuum Contributions To The Pion Radius, Phys. Rev. 178:2167-2169,1969; Character Analysis of U(N) and SU(N), J. Math. Phys. <u>10</u>, 2156 (1969); and The Calculation of the Irreducible Characters of the Symmetric Group in Terms of the Compound Characters, (Published as Blaha's Lemma in D. E. Knuth's book: *The Art of Computer Programming Vols. 1 – 4*).

In the early 1980's Blaha was also a pioneer in the development of UNIX for financial, scientific and Internet applications: benchmarked UNIX versions showing that block size was critical for UNIX performance, developing financial modeling software, starting database benchmarking comparison studies, developing Internet-like UNIX networking (1982) and developing a hybrid shell programming technique (1982) that was a precursor to the PERL programming language. He was also the manager of the AT&T ten-year future products development database. His work helped lead to commercial UNIX on computers such as Sun Micros, IBM AIX minis, and Apple computers.

In the 1980's he pioneered the development of PC Desktop Publishing on laser printers. and was nominated for three "Awards for Technical Excellence" in 1987 by PC Magazine for PC software products that he designed and developed.

Recently he has developed a theory of a Megaverse – containing universes of which our universe is one – with quantum particle-like properties based on the Wheeler-DeWitt equation of Quantum Gravity. He has developed a theory of a baryonic force, which had been conjectured many years ago, and estimated the strength of the force based on discrepancies in measurements of the gravitational constant G. This force, operative in D-dimensional space, can be used to escape from our universe in "uniships" which are the equivalent of the faster-than-light starships proposed in the author's earlier books. Thus travel to other universes, as well as to other stars is possible.

Blaha also considered the complexified Wheeler-DeWitt equation and showed that its limitation to real-valued coordinates and metrics generated a Cosmological Constant in the Einstein equations.

The author has also recently written a series of books on the serious problems of the United States and their solution as well as a book on the decline of Mankind that will follow from current social and genetic trends in Mankind.

In the past twelve years Dr. Blaha has written over 40 books on a wide range of topics. Some recent major works are: *From Asynchronous Logic to The Standard Model to Superflight to the Stars, All the Universe!, SuperCivilizations: Civilizations as Superorganisms, America's Future: an Islamic Surge, ISIS, al Qaeda, World Epidemics, Ukraine, Russia-China Pact, US Leadership Crisis,The Rises and Falls of Man – Destiny – 3000 AD: New Support for a Superorganism MACRO-THEORY of CIVILIZATIONS From CURRENT WORLD TRENDS and NEW Peruvian, Pre-Mayan, Mayan, Anatolian, and Early Egyptian Data, with a Projection to 3000 AD,* and *Mankind in Decline: Genetic Disasters, Human-Animal Hybrids, Overpopulation, Pollution, Global Warming, Food and Water Shortages, Desertification, Poverty, Rising Violence, Genocide, Epidemics, Wars, Leadership Failure.*

He has taught approximately 4,000 students in undergraduate, graduate, and postgraduate corporate education courses primarily in major universities, and large companies and government agencies.